Climate Change

Reasons To Worry Less

by Steve McGee

Foreword by Carla Deemer

ISBN: 9798543014547

Table of Contents

Foreword

"I have no idea," I reply to Steve's friends when they ask me what he's been doing with his retirement. "Perhaps he walks his dog. I know he goes to the library...", I trail off trying to dodge the question. For as long as I've known him, Steve has been pretty quiet about the things he does, preferring not to discuss projects until they've been completed. So when his friends do ask me this question, it is always in his absence and in invariably hushed tones, as though trying to extract some secret to which only I am privy.

Truthfully, though, I do not know what he does all day. This is why, when he placed the slender, purple book in my hands for the first time and my eyes focused on the glossy cover, finally recognizing the author to be the same Steve McGee whom I've known all these years, I exclaimed, "OH!" with genuine surprise. And then I read the title and let escape a groan accompanied by an involuntary eye-roll. *Oh. Climate change.* That *pesky subject. This is what he's been doing all day...* I caught myself having the familiar gut-reaction I am wont to have when confronted with anything claiming to refute the dangers of global warming. But why do I have that reaction? Why am I not receptive to this information, even if it's in the form of data and graphs?

My generation, the so-called millennials, had the deleterious effects of global warming drilled into us from a very early age. I can not remember a time when there wasn't some green initiative or other such cause being undertaken in the schools I attended, or seeing the hottest celebrities hosting extravagant galas to raise money to somehow reduce carbon emissions (while traveling the world in private jets). So that is why, when I let slip to a close friend that Steve and I occasionally argue about climate change, she responded, "Oh, you mean like what's the best way to fix it?" as though I couldn't possibly associate with someone who may have a different point of view on the matter altogether. If you believe the earth is in danger because of man-made pollution (which, if you're my age, is what you've been taught your whole life), of course you're going to argue in favor of change. After all, what's more important than saving the world?

With nearly 20% of US adults suffering from some form of anxiety disorder (https://www.nimh.nih.gov/health/statistics/any-anxiety-disorder.shtml) and 7.1% of US adults with diagnosed major depressive disorder (https://www.nimh.nih.gov/health/statistics/major-depression.shtml), perhaps one would be clever to curb the "we're all doomed and the world is ending" mentality and possibly alleviate some of the anxiety caused by that mantra. We are inundated with constant reminders that the world as we know it is ending, that forest fires are going to engulf us all while we are simultaneously being drowned by the ever-rising oceans, and that there soon won't be any trees left for our children to enjoy. Maybe it's time we take a step back and look at the data. Granted, depictions of the earth literally melting are more sensational that numbers on graphs, but perhaps it is the plain data we should look at to form a clearer understanding of what is happening with the climate, if a clear understanding can be had at all. Mr. McGee has made 49 data based arguments to ease our fears that the world will melt away. Not in denial of global warming, he has shed light on what appear to be some of the grayer areas in the climate change debate. These are the types of arguments we need to focus on - level-headed, unemotional data.

Let's all just Chill.

- Carla Deemer

Introduction

When writing about climate change, one must consider the background of the reader. Atmospheric processes can be complex, detailed and wide ranging. What follows does not contain complex physical formulas, but it is still not for everyone. The following work presumes the reader has a basic understanding of climate change and an affinity for charts and graphs.

The popular concept of climate change is that the increase of certain gasses including carbon dioxide, reduce outgoing energy from earth, raising mean temperature and perhaps causing indirect impacts. Those familiar with the ideas described in this one sentence are more likely to receive the ideas which follow.

The thesis of this brief book embraces the principles of climate change as demonstrable, true, and well understood. This work diverges from popular understanding in considering the bounds and context of climate change, the observed data which contradicts some presumed changes, the inclusion of climate change benefits, and the biases which can pervade any subject including climate change.

Is climate change unbounded and unlimited? Is climate change completely anthropogenic? Is climate change completely adverse? Could benefits of climate change exceed detriments? Are the majority of presumed adverse climate change events becoming more frequent? Can climate models accurately predict the state of the atmosphere? What is the role of psychological bias in conceiving of climate change? My hope is for the reader to consider these questions while reading this text to continue to do so as part of their assessment of climate change.

Warming Less Than Expected

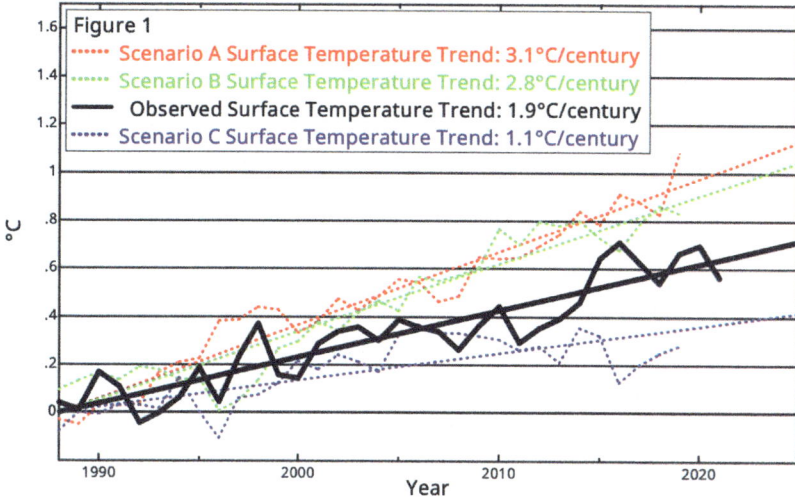

Figure 1

In 1988, the United States Senate heard testimony from Dr. James Hansen of the National Aeronautics and Space Administration (NASA) Goddard Institute for Space Studies (GISS).[1.1] The testimony was based on climate models of three scenarios of different greenhouse gas emissions. Scenario A was called "business as usual" and included high levels of greenhouse gas emissions. Scenario C was called "draconian emission cuts" in which greenhouse gas emissions completely ceased in the year 2000. Scenario B was an intermediate scenario.

Figure 1 depicts the global mean surface temperature anomalies and trends from the three NASA-GISS scenarios with dotted lines.[1.2] Figure 1 also depicts observations from the National Oceanic and Atmospheric Administration (NOAA) with a thick solid line.[1.3] Consistent with the theory that increased radiative forcing will cause global warming, the observed temperature record indeed indicates a warming trend. But since the 1988 testimony, observed temperature trends have been less than those of the expected intermediate scenario.

Warming Not Rapidly Accelerating

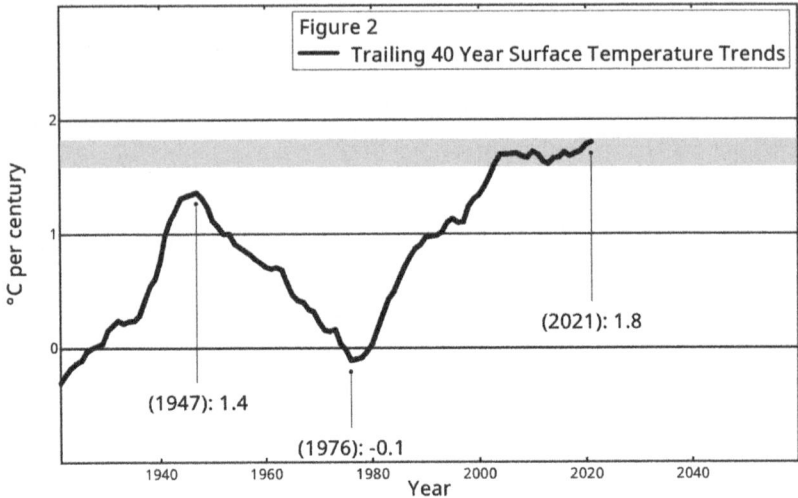

Figure 2
Trailing 40 Year Surface Temperature Trends

(1947): 1.4

(1976): -0.1

(2021): 1.8

Depicted in Figure 2 is the time series of trailing forty year surface temperature trends.[2.1] In 1947, the forty year warming trend reached 1.4°C per century. By 1976, this had changed to a cooling trend. The latest forty year trend indicates a warming trend of about 1.8°C per century. This rate has been relatively stable for the past two decades.

Data from the most recent years do indicate a slight increase of trends. As we shall see below, this recent small increase may be due to known factors. The timing of the El Chichón and Mount Pinatubo volcanic eruptions tends to inflate these trends. This effect will begin to decrease in future years. There are also indications that earth has absorbed more sunshine than is normal since the year 2000. Whether this increase represents a new normal or is subject to reversion to the mean will be important to monitor.

Warming At A Low Scenario

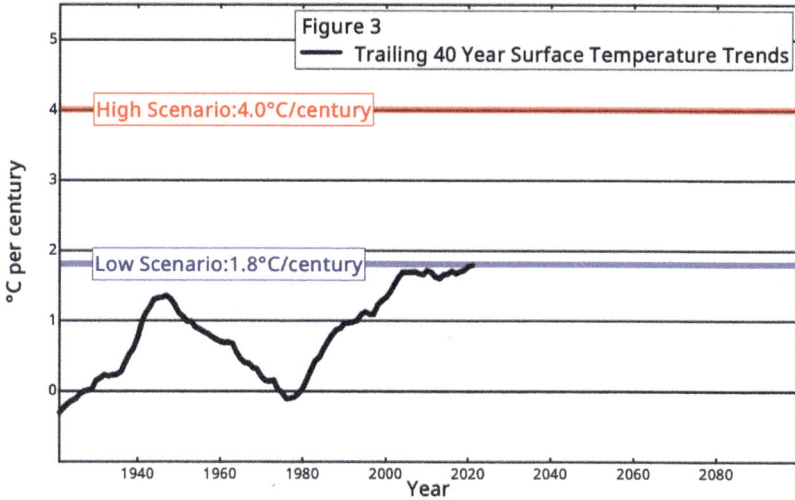

Figure 3 again includes surface temperature trends.[3.1] This time, the past trends are in the context of projected scenarios.[3.2] The scenarios of this figure are from the 2007 Synthesis Report of the Inter-governmental Panel on Climate Change (IPCC), which estimated rates of warming for the remainder of the twenty-first century. In this report, and as indicated in Figure 3, the best estimate for a high scenario was 4.0°C per century. The best estimate for a low scenario was 1.8°C per century. The most recent forty year trend is approximately this same 1.8°C per century low scenario rate. Contrary to popular understanding that global warming is extreme, the observed surface temperature trend is approximately at the best estimate for the low scenario.

Forcing Less Than Expected

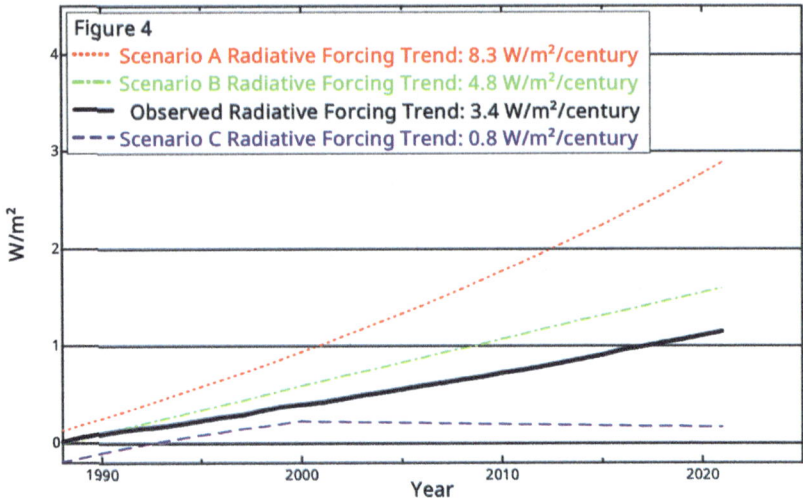

Figure 4
- Scenario A Radiative Forcing Trend: 8.3 W/m²/century
- Scenario B Radiative Forcing Trend: 4.8 W/m²/century
- Observed Radiative Forcing Trend: 3.4 W/m²/century
- Scenario C Radiative Forcing Trend: 0.8 W/m²/century

As described in the background section, increased radiative forcing is likely to continue to produce global warming. Figure 4 depicts the three modeled scenarios of the 1988 NASA-GISS testimony in terms of the cumulative radiative forcing.[4.1] Radiative forcing is related to the amount of energy retained by the atmosphere due to the increase of gasses such as carbon dioxide. Also shown is the cumulative radiative forcing estimated by NOAA to have actually occurred.[4.2] Estimated radiative forcing trends since 1988 have been at rates less than those of the expected Scenario B.

Forcing Rates Decreasing

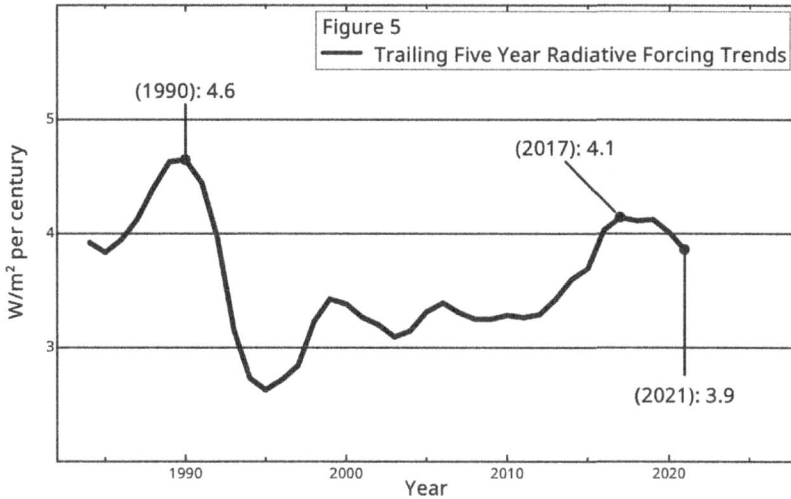

Figure 5 depicts the rates of change of radiative forcing from greenhouse gasses.[5.1] The values are plotted as trends over five years to smooth out shorter term fluctuations. Radiative forcing occurs from changes in greenhouse gasses, including carbon dioxide, methane, nitrous oxide and chlorofluorocarbons. The peak rates of radiative forcing occurred around 1990. The subsequent decline of radiative forcing rates was due to the decrease of chlorofluorocarbons. The increased rate of radiative forcing from around 2001 coincides with increased carbon dioxide from the economic growth and development of China following permanent normal trade relations with the US. The five year radiative forcing rates reached a secondary maxima in 2017 and have decreased in each of the subsequent years since. As we shall see in the next section, there are reasons to believe that rates of radiative forcing will continue to decrease.

Decreasing CO$_2$ Emission Rates

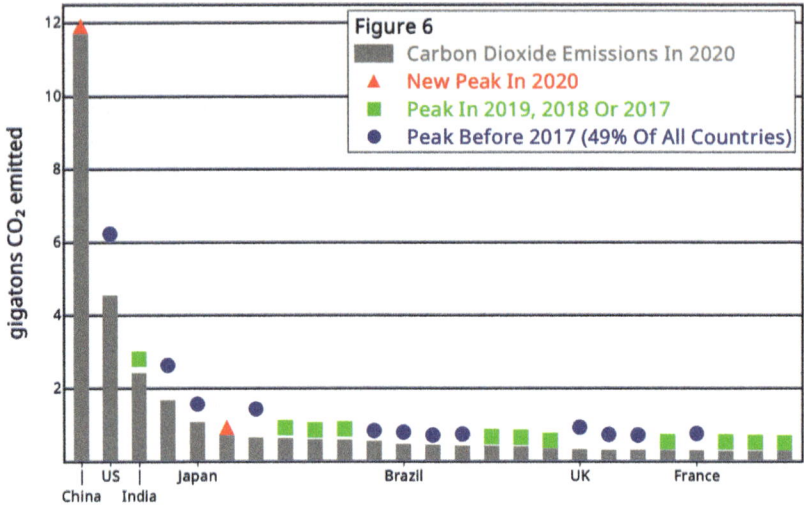

Figure 6

- ▲ New Peak In 2020
- ■ Peak In 2019, 2018 Or 2017
- ● Peak Before 2017 (49% Of All Countries)

Carbon Dioxide Emissions In 2020

Figure 6 is a graph of carbon dioxide emissions from the highest emitting countries.[6.1] Bars represent the most recent emissions, triangles indicate a peak in the most recent year, squares indicate a peak in the previous three years, and circles indicate more than three years past peak. Around half of all countries had peak emissions occurring more than three years prior to the most recent year.

Decreasing Per Capita CO$_2$ Rates

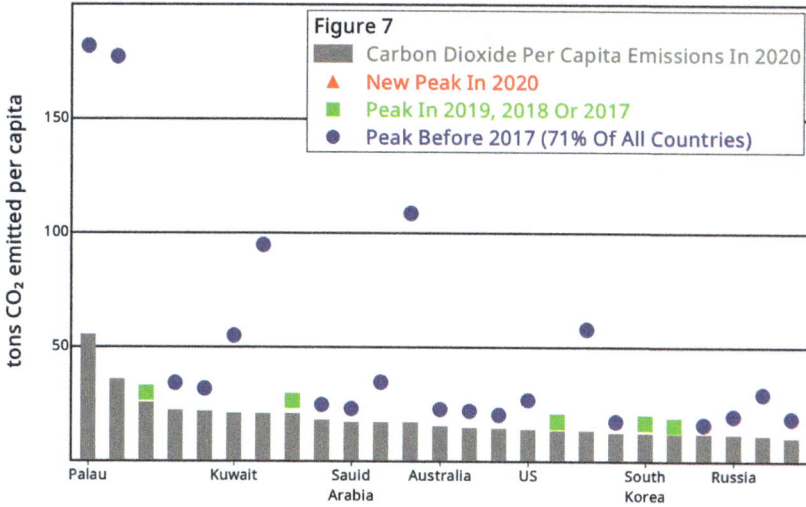

Figure 7
- Carbon Dioxide Per Capita Emissions In 2020
- ▲ New Peak In 2020
- ■ Peak In 2019, 2018 Or 2017
- ● Peak Before 2017 (71% Of All Countries)

Figure 7 is a graph of per capita carbon dioxide emissions for the highest per capita emitting countries.[7.1] Bars represent the most recent per capita emissions, triangles indicate a peak in the most recent year, squares indicate a peak in the previous three years, and circles indicate more than three years past peak. Most countries had peak per capita emissions more than three years prior to the most recent year.

Decreasing Global Per Capita CO$_2$

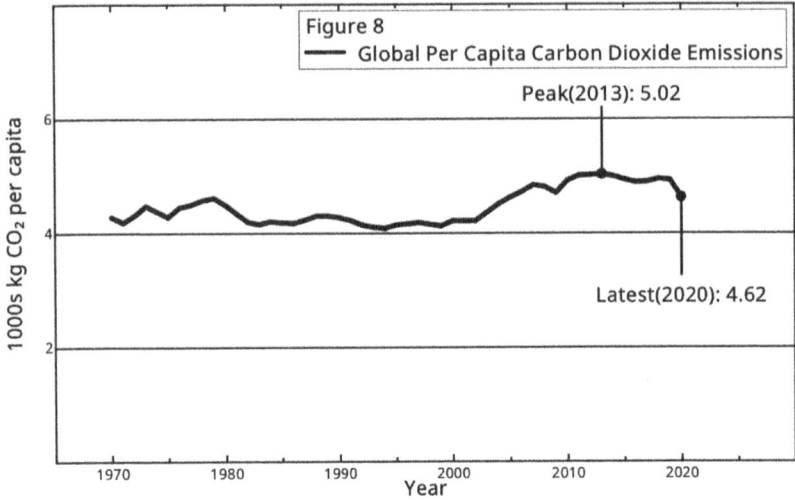

Figure 8
Global Per Capita Carbon Dioxide Emissions

Peak(2013): 5.02

Latest(2020): 4.62

Y-axis: 1000s kg CO$_2$ per capita
X-axis: Year (1970, 1980, 1990, 2000, 2010, 2020)

Since most countries have per capita carbon dioxide emissions which are past peak, it should not be surprising that Figure 8 also indicates that global per capita carbon dioxide emissions are also past peak.[8.1] To be sure, decreases from peak have occurred previously. The global measure reflects the state of development of each component nation. While developing, nations tend to emit more carbon dioxide per capita because infrastructure development requires energy use. However, once developed, nations may emit less carbon dioxide because the newly developed infrastructure increases efficiency.

Fertility Rates Are Decreasing

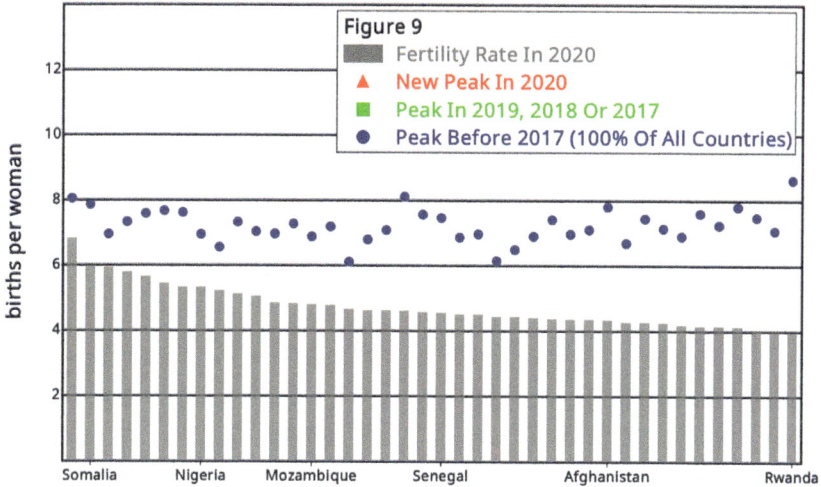

Figure 9
- Fertility Rate In 2020
- ▲ New Peak In 2020
- ■ Peak In 2019, 2018 Or 2017
- ● Peak Before 2017 (100% Of All Countries)

Population is determined by death rates and birth rates. The number of births per woman over a lifetime is known as the total fertility rate. Figure 9 is a chart of recent fertility levels compared to the peak level of fertility for the forty most fertile countries.[9.1] Rather remarkably, total fertility rates are declining for 100% all countries of earth!

George Friedman points out that in agrarian societies, young children can be economically productive, collecting food and tending to animals, while in modern societies, a child going to graduate school might require expensive education without being economically productive until age twenty-six.[9.2] In countries where children are productive at a young age, there is increased economic incentive toward having more children. In countries where children require lengthy and costly education, there is economic incentive toward having fewer children or perhaps no children. The incentives associated with economic development can explain much of the falling fertility rates worldwide. Falling fertility means fewer births which ultimately limits carbon dioxide emissions.

Sub-Replacement Fertility Rates

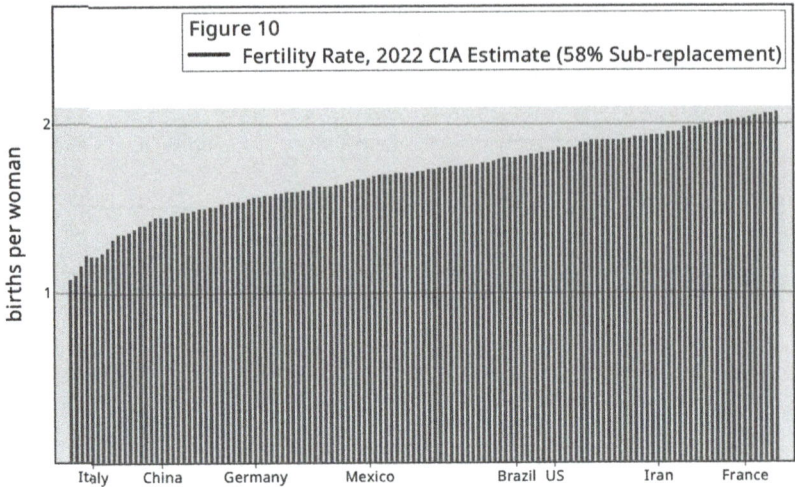

Figure 10
— Fertility Rate, 2022 CIA Estimate (58% Sub-replacement)

An important value of total fertility rate to consider is 2.1 births per woman, which is approximately the average number of children born per woman to replace herself and her partner and result in a stable population size. This value is known as the replacement fertility rate. This number is slightly greater than two because mortality means a certain number of females perish before reaching child bearing years. The precise replacement rate can vary slightly from country to country based on longevity and early mortality differences. Figure 10 is a chart representing the fertility rates of countries with lower than replacement fertility.[10.1] More than half of all countries now have fertility rates which are lower than the replacement rate.

Decreasing Global Fertility Rates

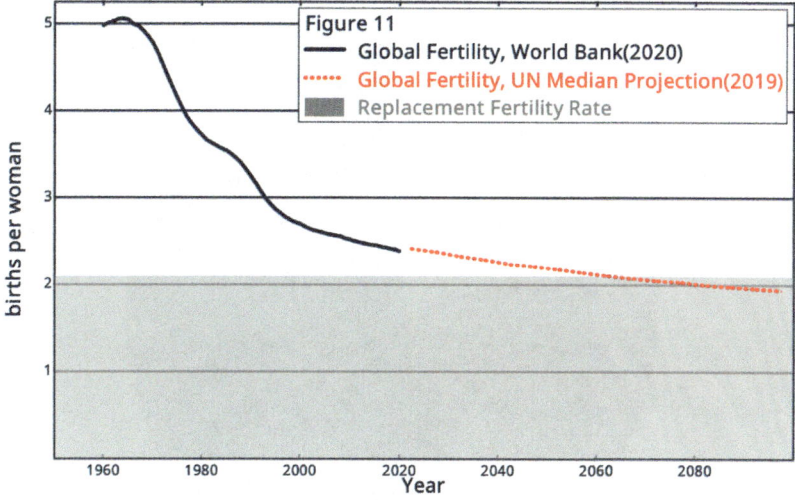

Figure 11

- Global Fertility, World Bank(2020)
- Global Fertility, UN Median Projection(2019)
- Replacement Fertility Rate

Figure 11 is a plot of global fertility. The solid line indicates past fertility rates.[11.1] The dotted line indicates the median predicted future fertility rates.[11.2] The shaded area represents the replacement fertility rate. Fertility rates less than the replacement rate imply eventual falling population rates. Fertility rates are approaching the replacement rate globally. The latest estimate for fertility rates are slightly less than the median predicted future rate. The median predicted trend of fertility from the United Nations indicates less than replacement global fertility rates around the year 2064, with decreasing population soon thereafter. Extrapolation of the trend of past estimates of fertility suggests replacement rates of fertility will occur somewhat sooner than the median projection. In addition to falling rates of per capita carbon dioxide emissions, the looming decrease of population will tend to compound decreasing carbon dioxide emissions.

CO$_2$ Emissions & Low Fertility

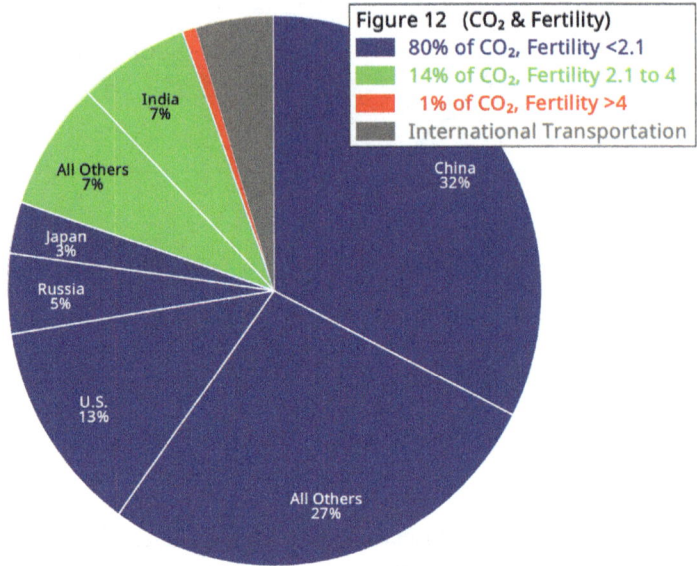

Figure 12 (CO$_2$ & Fertility)
■ 80% of CO$_2$, Fertility <2.1
■ 14% of CO$_2$, Fertility 2.1 to 4
■ 1% of CO$_2$, Fertility >4
■ International Transportation

China 32%

India 7%

All Others 7%

Japan 3%

Russia 5%

U.S. 13%

All Others 27%

In Figure 12, the size of each pie slice represents the percentage of global carbon dioxide emissions from each country.[12.1] The shading of each pie slice represents the fertility rate of each country.[12.2] Around 80% of all carbon dioxide emissions are from countries with lower than replacement fertility rates. The primary implication of Figure 12 is that the inevitable decreased population from low fertility will reduce carbon dioxide emissions through fewer emitters in the future.

While the main point of Figure 12 is about decreased carbon dioxide emissions because of looming population decline, it's worth considering the effect of economic development on both fertility and carbon dioxide emissions. Limiting population became a concern during previous decades when fertility rates were very high. Many thought that limiting economic development was the solution in order to limit population. But the economic status and fertility of the countries depicted in Figure 9, Figure 10, and Figure 12 indicates just the opposite. The countries with the lowest fertility rates are the most economically developed. Apparently, it is not the limitation of economic development, but rather the fulfillment of economic development that limits population. To be sure, this economic development has taken place largely by using readily available fossil fuels. But by reducing fertility and by increasing efficiency, economic development may create a negative feedback. Increased carbon dioxide emissions from economic development in the short term may lead to decreasing carbon dioxide emissions in the longer term.

CO$_2$ Uptake Has Increased

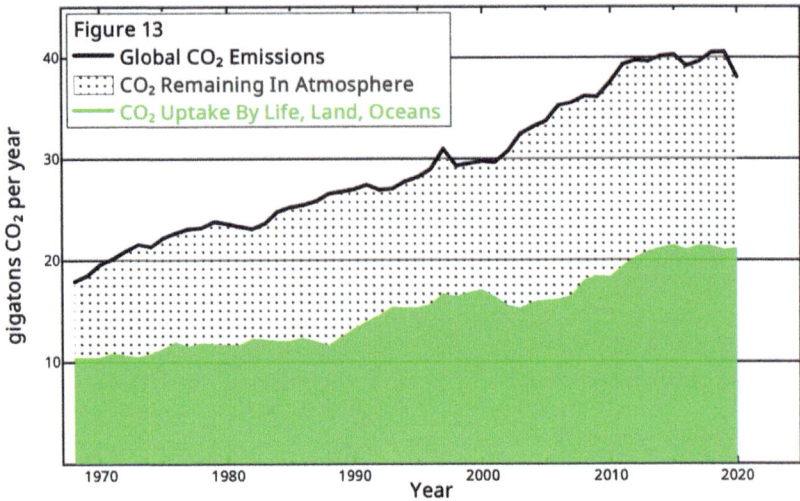

Figure 13
— Global CO$_2$ Emissions
⬚ CO$_2$ Remaining In Atmosphere
— CO$_2$ Uptake By Life, Land, Oceans

The solid line of Figure 13 indicates the level of carbon dioxide emissions from human activity. The dotted area indicates the portion of human released carbon dioxide remaining in the atmosphere. The solid area represents the portion of human released carbon dioxide leaving the atmosphere and taken up by life, land, and oceans.[13.1, 13.2] The plotted values represent ten year means to smooth out shorter term fluctuations.

As we shall see in following reasons, there is evidence that much of this uptake of carbon dioxide is going toward increased life on earth. And as one can see, the amount of carbon dioxide taken up has increased over the past half century. An implication of increased uptake is that carbon dioxide emissions needn't cease completely in order to cease accumulation of carbon dioxide in the atmosphere.

Temperature Changes Reverse

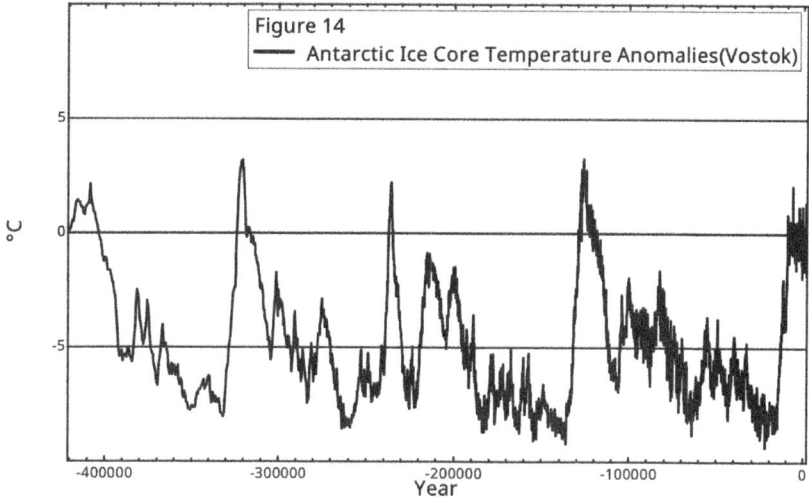

Figure 14

Figure 14 is a plot of the proxy temperature change from Antarctica for more than four hundred thousand years.[14.1], [14.2], [14.3], [14.4], [14.5] The proxy is not a direct measurement of temperature by thermometer, but rather is estimated from changes of isotopes in the deep layers of ice which are correlated with temperature. This relationship is not perfect, but provides an estimate of past temperatures near the regions where snow and ice accumulate. Of note are the periods for which the temperature difference with respect to modern temperatures were greater than zero. These periods of warmth, known as interglacials, contrast with the slow descent of temperature for much longer periods which are known as glacials, or more commonly, the ice ages. The ice ages lasted for relatively much longer periods than the interglacials. Inter-glacial periods have followed every glacial period. In this regard, climate change has always reversed.

Some Variation Is Normal

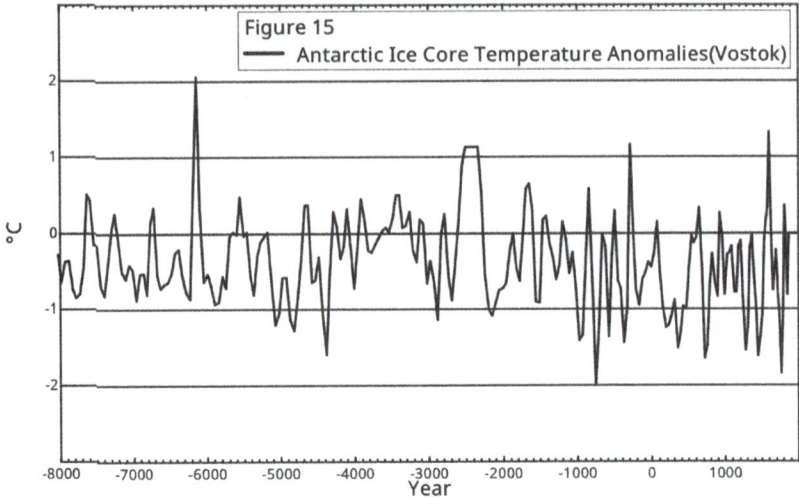

Figure 15
— Antarctic Ice Core Temperature Anomalies(Vostok)

A shorter time period of the Antarctic proxy temperature record is depicted in Figure 15.[15.1, 15.2, 15.3, 15.4, 15.5] Plotted are the proxy temperature differences for the last ten thousand years, which covers much of the period of human civilization. While there is uncertainty with the proxy estimates, and this proxy is from Antarctica only, the record indicates natural fluctuations of roughly plus or minus 1°C. Observed century rates of global warming, as depicted in Figure 2, are only slightly higher than this. Some temperature fluctuation is normal and the estimated global warming caused by humans is so far still comparable to this level of natural fluctuation.

Warmer, Snowier Greenland

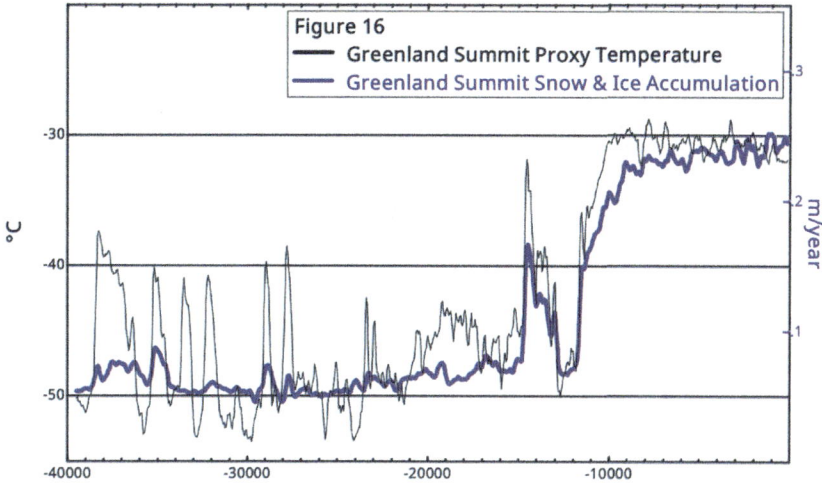

Figure 16
— Greenland Summit Proxy Temperature
— Greenland Summit Snow & Ice Accumulation

Figure 16 depicts more proxy data from ice cores, this time not from Antarctica, but from the Greenland Summit.[16.1, 16.2] The thin line represents not the temperature change, but rather the absolute average temperature. The thick line is a plot of the estimated snow and ice accumulation expressed in meters per year. The estimate of ice accumulation is somewhat uncertain due to compression and deformation of the snow and ice layers. One may note the increase of temperatures from around minus 50°C to around minus 30°C some ten thousand years ago. This increase in temperatures corresponds with the end of the last ice age. While ice sheets elsewhere did decrease as temperatures increased, ice accumulation increased over the Greenland Summit, where the average annual temperature of minus 30°C is still well below freezing.

Modern snow accumulation over the higher terrain of Greenland is exemplified by the remarkable story of the *Glacier Girl*.[16.3, 16.4] In 1942, during World War II, US aircraft returning from the United Kingdom made emergency landings on the higher terrain of Greenland. All the crew were rescued, but the air craft were left behind. When a 1992 expedition found some of the planes, they were excavated from a depth of approximately

82 meters of snow and ice. And recently, a 2018 expedition found another of the planes, this time under 91 meters of snow and ice.[16,5] The total mass of Greenland ice depends on ice shed to sea and lower elevation melting in addition to accumulation. It remains to be seen whether global warming acts analogously, but over the previous glacial cycle, snow accumulation increased as the temperatures over Greenland increased.

Migration & Climate Change

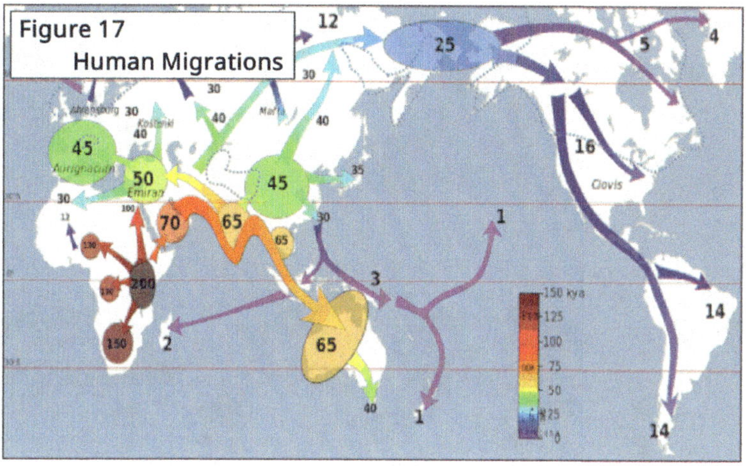

Figure 17
Human Migrations

Figure 17 is a map of theorized ancient migrations based on human remains.[17.1] The numeric values on this map indicate the age of these remains in thousands of years. It appears that human beings spread across the globe to every climate on earth. We consider modern climate change over time, but these migrations remind us that during evolution, humans self imposed climate change by moving in space to different climates. Fortunately, these preceding humans were not frightened of climate change they encountered on their migrations and their legacy is written in our collective human DNA, making us who we are today. We humans are evolved for climate change because our ancestors changed the climates in which they lived by migrating.

Warming & Paleo-Vegetation

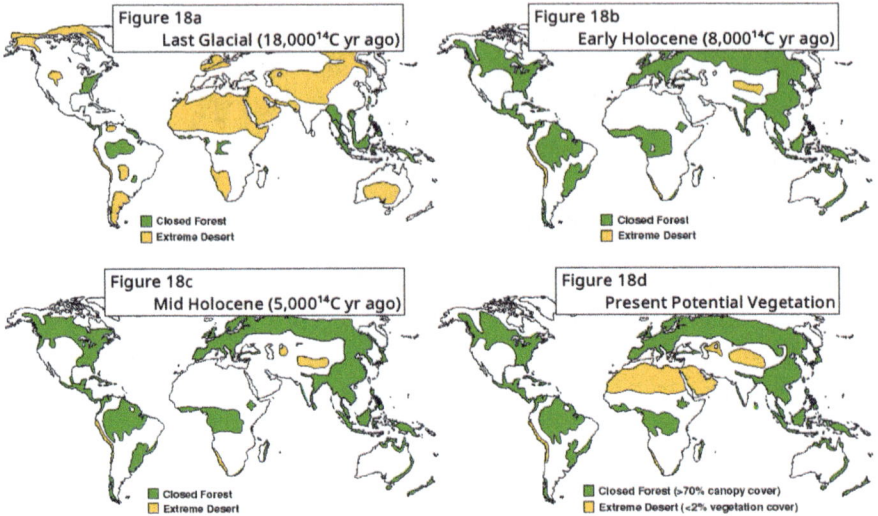

Figure 18a
Last Glacial (18,000¹⁴C yr ago)

Closed Forest
Extreme Desert

Figure 18b
Early Holocene (8,000¹⁴C yr ago)

Closed Forest
Extreme Desert

Figure 18c
Mid Holocene (5,000¹⁴C yr ago)

Closed Forest
Extreme Desert

Figure 18d
Present Potential Vegetation

Closed Forest (>70% canopy cover)
Extreme Desert (<2% vegetation cover)

Figures 18 a,b,c & d contain estimates of past vegetation from fossil records.[18.1, 18.2] Figure 18a contains a global vegetation category map for the most recent ice age. Figures 18b and 18c contain global vegetation maps for the peak warmth of the interglacial period. Figure 18d contains a global vegetation category map for present vegetation. These categorical maps are based in large part on numerous studies of pollen and plant fossil records. The categories are of closed forest, shaded in dark green, and of extreme desert, shaded in gold. There is of course some uncertainty about the maps due to limited resolution and confounding variables. However, the results of the analysis are quite contrary to popular understanding. According to this analysis, areas of extreme desert were much more prevalent during the cold of the most recent ice age. Areas of closed forest were much more prevalent during the warmth of the subsequent interglacial period, including recent times. Of course, factors other than temperature also changed during these periods. Also, the temperature changes of the glacial cycles were not seasonally uniform, so the analog is not perfect. But over the most recent glacial cycle, vegetation appears to increase with increased global mean annual temperature.

Nature Will Melt Arctic Ice

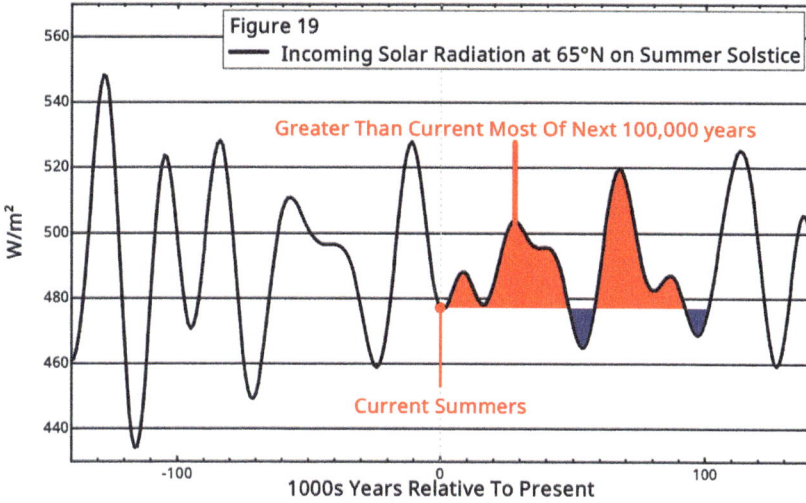

Figure 19
— Incoming Solar Radiation at 65°N on Summer Solstice

Greater Than Current Most Of Next 100,000 years

Current Summers

W/m²

1000s Years Relative To Present

The ice ages evident in the proxy temperature records from ice cores are thought to occur because of variations of earth's orbit.[19.1, 19.2] Earth rotates about a tilted axis and the tilt of this axis tends to wobble. Also, earth's orbit around the Sun varies from more nearly circular to more elliptical. Further, the elliptical orbit can vary between putting the Northern Hemisphere closest to the sun during winter or during summer. Such variations may determine whether snow and ice melt or accumulate. Theses variations are named the Milanković Cycles after the Serbian mathematician Milutin Milanković who identified them.[19.3] The orbital variations can determine the variation of sunshine at the Northern Polar regions. There are other internal variations of temperature and precipitation. Further, it may take thousands of years for ice to accumulate to a kilometer or so where ice is further protected by the lower temperatures of that altitude. Because of these and other factors, glacial cycles do not perfectly match orbital variations. But the highly predictable orbital variations are still thought to be the most important determinant of the ice ages.

Figure 19 represents the amount of summer sunshine at 65° North

Latitude. The past peaks and troughs of this value roughly correlate with glacial retreat and advance. For most of the next one hundred thousand years, there will be greater summer sunshine at the Northern Polar regions. Independent of humans, this natural increased summer sunshine will provide direct energy to melt susceptible Arctic ice.

Some Sea Ice Loss From Wind

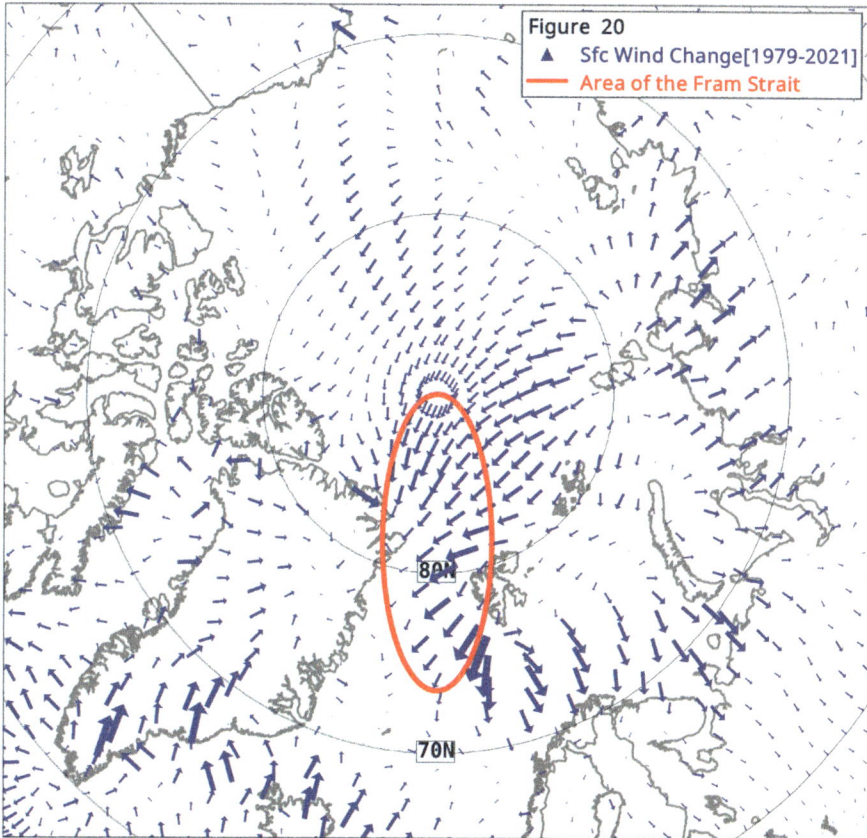

Figure 20
▲ Sfc Wind Change[1979-2021]
— Area of the Fram Strait

One change thought to occur with global warming is the loss of sea ice in the Arctic Ocean. Global warming may directly reduce Arctic sea ice by melting in place from increased temperatures. Arctic sea ice is also mobile. It tends to slowly move with the winds. Winds that tend to blow in a circle around the Arctic tend to allow ice to accumulate. One such motion is known as the Beaufort Gyre. Winds that tend to blow outward through the narrow gaps of the Arctic Ocean tend to decrease sea ice. One such current is known as the Transpolar Drift,

Figure 20 presents a map of the changes of winds based on reanalysis of data.[20.1] Reanalyses combine past observations to create gridded data sets in a physically consistent manner. The actual observations may be quite sparse and there is uncertainty associated with reanalysis. The dashed oval indicates a region where winds changed to blow sea ice away from the Arctic and toward lower latitudes. This region is known as the Fram Strait. According to the reanalysis, at least part of Arctic sea ice loss appears to be from fluctuation of winds blowing more than average away from the Arctic and toward lower latitudes.

Temperature Trends & Volcanoes

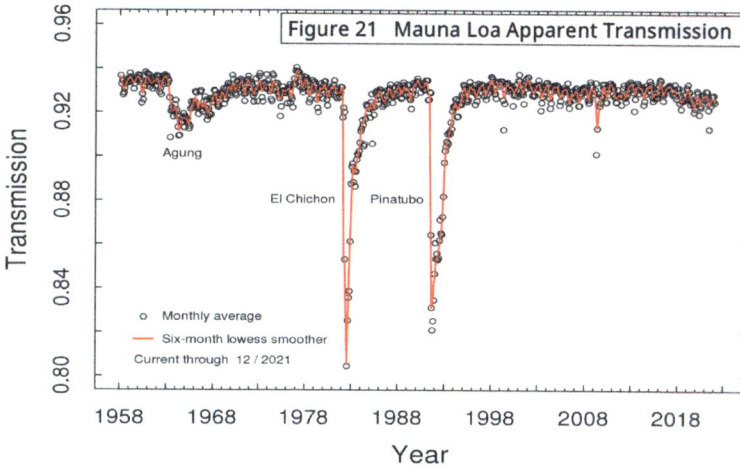

Figure 21   Mauna Loa Apparent Transmission

Radiative forcing tends to focus on changes of the outgoing energy from earth. Natural changes of the amounts of energy earth absorbs from sunshine can also alter radiative forcing. Occasionally, volcanoes erupt and send large amounts of dust and sulfur dioxide high into the atmosphere. This dust reflects away sunshine and reduces the amount of incoming energy available to warm the earth.

Figure 21 depicts an estimate of the amount of solar energy transmitted through the upper atmosphere based on daily observations from the Mauna Loa Observatory in Hawaii.[21.1] The eruption of El Chichón in 1982 and the eruption of Mount Pinatubo in 1991 are evident in the decrease of atmospheric transmission. Because there have been no similar eruptions since, trends of radiative balance since the 1970s, 1980s or 1990s may include a positive trend of incoming sunshine only because of the timing of the eruptions. Radiative forcing from increased carbon dioxide remains a substantial cause of warming. But at least some portion of the observed warming trend from the late 1970s may be from the timing of volcanic events.

Warming From Absorbed Sunshine

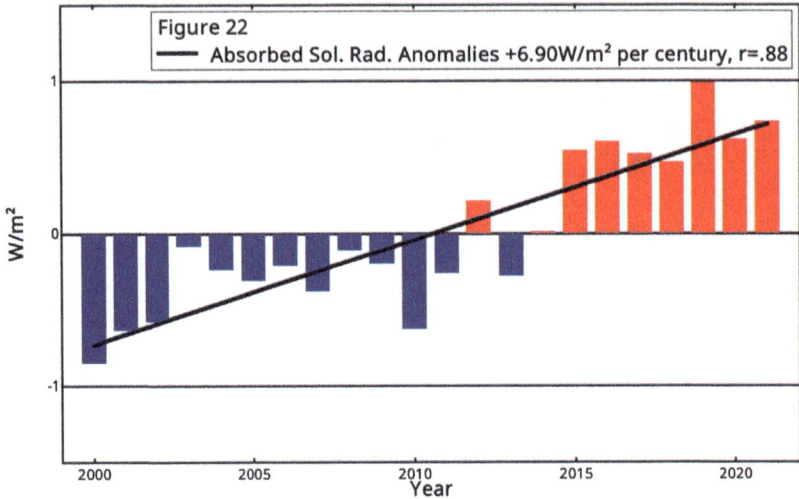

Figure 22
—— Absorbed Sol. Rad. Anomalies +6.90W/m² per century, r=.88

In addition to the dusty volcanic eruptions discussed above, a factor determining how much sunshine is reflected away before it can heat the earth is cloudiness. The surface of the earth reflects away some sunshine, but clouds reflect much more than the surface does.[22.1] Because they are relatively small, transient, and irregular, clouds are not particularly well observed. Clouds can reflect more sunshine to space by being more prevalent, by having more droplets within them, by being lower based, and by a number of other factors. Because of these factors, there is uncertainty with estimates of reflection.

Figure 22 presents the change in the amount of sunshine absorbed by earth.[22.2] This value is calculated by subtracting the amount of reflected sunshine from the amount of sunshine incident on earth as observed by orbiting satellites. While this duration is relatively short, it is apparent that the maximum years of absorbed sunshine are all since 2015. As with other climatological data sets, there is much uncertainty. However, these estimates indicate that some of the warming observed since 2000 may be due in large part to increased absorption of sunshine.

Sea Level Rise From Groundwater

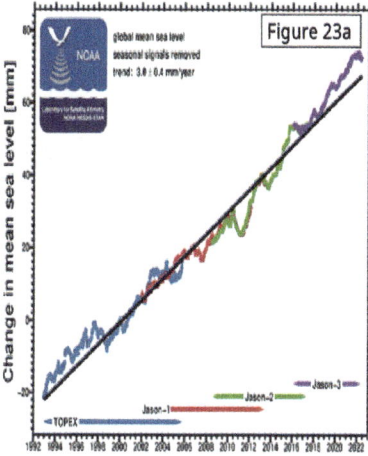

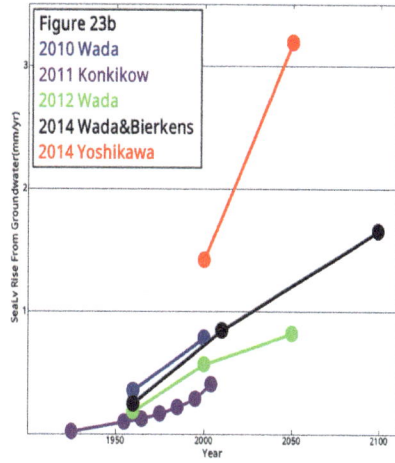

Figure 23a

Figure 23b
2010 Wada
2011 Konkikow
2012 Wada
2014 Wada&Bierkens
2014 Yoshikawa

Sea level is expected to rise with global warming. Warming of the ocean waters causes thermal expansion of those waters which leads to rising sea levels. Figure 23a indicates that, indeed, the satellite based global estimate of sea level is increasing.[23.1] The estimate of the global trend of sea level rise is about 3 millimeters per year.

At least part of global sea level rise may be due to factors other than global warming. Groundwater pumped from wells and used by humans eventually reaches the oceans. Groundwater use is evidently a large and accelerating portion of sea level rise. Figure 23b depicts estimated past and predicted future rates of sea level rise attributable to groundwater use.[23.2, 23.3, 23.4, 23.5, 23.6, 23.7] There is a lot of variation from the estimates indicating uncertainty. The same arguments of fertility decrease should eventually apply to ground water, though by far the largest use of groundwater is for agricultural crop irrigation. According to research, at least some portion of observed sea level rise is not due to warming, but rather is due to groundwater use. Within coming decades, one of the estimates puts sea level rise from groundwater at a rate greater than the currently observed rate of sea level rise from all causes.

US Deaths Peak During Winter

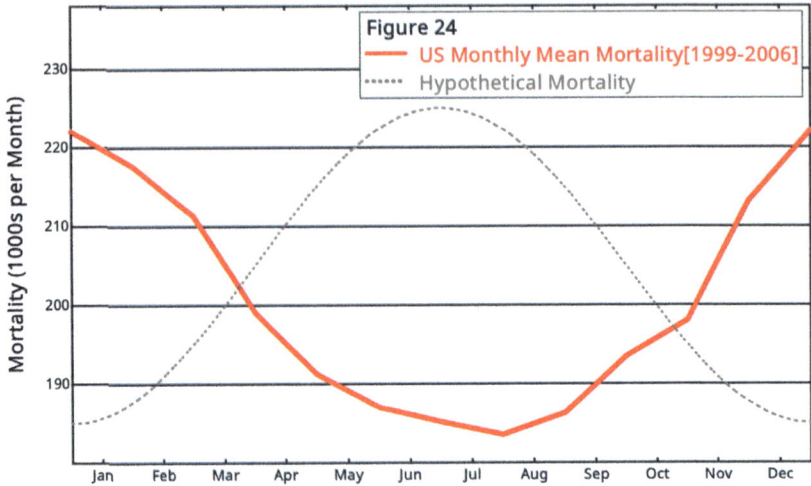

Figure 24
US Monthly Mean Mortality[1999-2006]
Hypothetical Mortality

A popular idea about global warming is that it is deadly. According to this idea, higher global average temperatures may lead to more human deaths. One available though imperfect analogue of this idea is the variation of mortality by season. Consider how human mortality varies with the much larger seasonal variation of temperature.

The dotted line of Figure 24 depicts what we might expect seasonally if temperature was a major determinant of death. If higher temperatures were a significant factor to mortality, we would expect more deaths during summer and fewer deaths during winter. The solid line of Figure 24 depicts the monthly average all cause mortality in the US for the years 1999 through 2006.[24.1] In near complete contrast to the expectations, significantly more people perish during the cold of winter, and significantly fewer people perish during the heat of summer. Of course, mortality has many factors, so it would be short sighted to ascribe mortality completely to low temperatures. While extra-tropical temperature varies with season, so too does sunlight. The production of vitamin D, the production of the vasodilator nitric oxide, and the production of hormones such as melatonin and serotonin, apparently all depend on sunshine.[24.2, 24.3, 24.4]

Rather than temperature, solar variation alone could account for the pattern of mortality indicated. But we may consider that warming could decrease mortality, warming could have little to no effect on mortality, or warming could increase mortality. Since seasonal mortality is anti-correlated with temperature, even if warming were to increase mortality, the effect would have to be comparatively small. Given the distribution, it would appear more likely that seasonal warming reduces mortality.

US Deaths From Cold & Heat

Figure 25 US Annual Weather Deaths[2006-2010]

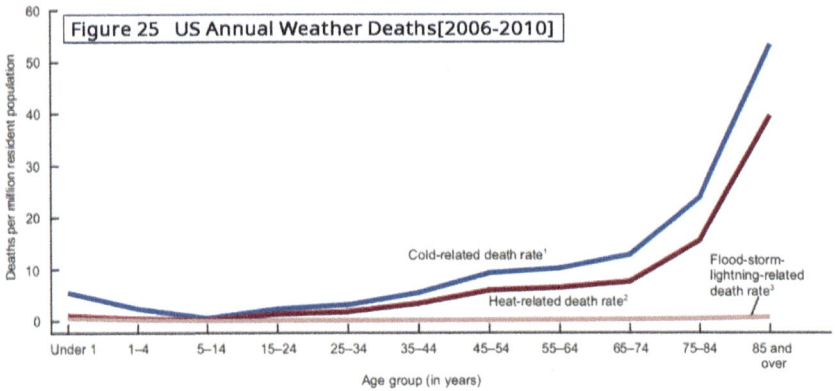

[1]Deaths attributed to exposure to excessive natural cold (X31) (underlying or contributing cause of death or both), to hypothermia (T68) (contributing cause of death), or to both, according to the International Classification of Diseases, 10th Revision.
[2]Deaths attributed to exposure to excessive natural heat (X30) (underlying or contributing cause of death or both), to heat stroke or sunstroke (T67) (contributing cause of death), or to both, according to the International Classification of Diseases, 10th Revision.
[3]Deaths attributed to floods (X38), cataclysmic storms (X37), or lightning (X33) (underlying or contributing cause of death or both), according to the International Classification of Diseases, 10th Revision.
SOURCE: CDC/NCHS, National Vital Statistics System, 2006-2010.

Seasonal mortality drops during the heat of summer and rises during the cold of winter. While this is true for the average temperature variation, there are a certain number of deaths each year from extreme temperatures. One may die from hypothermia and one may also die from heat stroke.

Figure 25 depicts the annual mortality rate from weather events in the US for the years 2006 through 2010.[25.1] The events are categorized as cold-related, heat-related, or floods-storm-lightning related. While they are often violent and dramatic and the deaths are all too tragic, floods, storms and lightning cause relatively few deaths. With respect to extreme temperature, cold related deaths exceed heat related deaths.

Weather Deaths Versus All Deaths

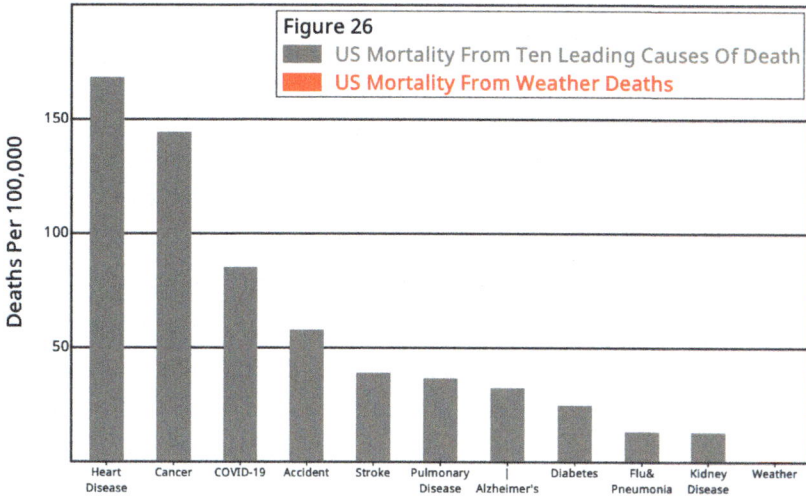

Figure 26

US Mortality From Ten Leading Causes Of Death

US Mortality From Weather Deaths

Figure 26 presents mortality rates from the ten leading causes of death in the US compared with mortality rates from weather derived from a Centers for Disease Control report.[26.1] The rate of deaths attributable to weather appears negligibly small in comparison.[26.2] Presumably, some number of deaths from weather have always occurred. Because heating, air conditioning and weather warnings have increased while the time spent outdoors has decreased, it is likely that weather deaths have decreased over the last century or so. There is no reproducible experiment to indicate whether a century of climate change will cause weather related mortality rates to increase, decrease, or stay about the same. But even if climate change were to increase weather mortality tenfold, it would remain a small fraction of mortality from other causes.

Preventable Deaths & Weather

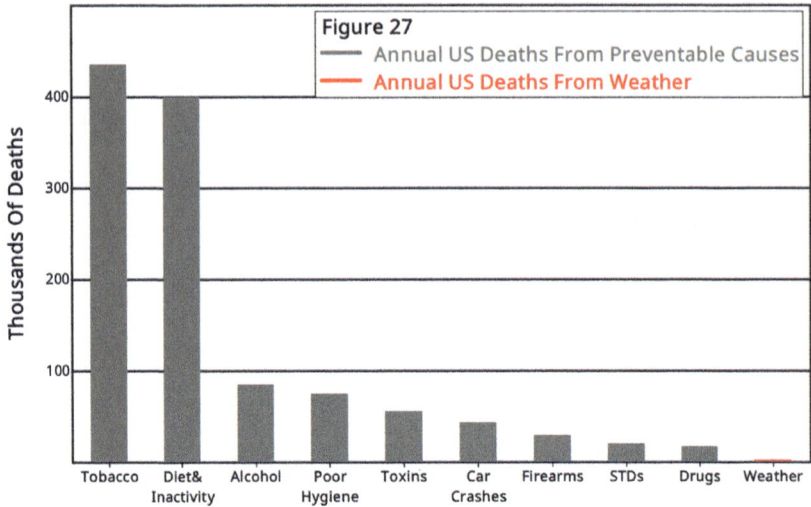

Figure 27
Annual US Deaths From Preventable Causes
Annual US Deaths From Weather

Figure 27 depicts total weather deaths in the context of estimated deaths from preventable risk factors.[27.1, 27.2] The figures are necessarily estimates with uncertainty. If one is worried about deaths from climate change but still smokes tobacco, it appears one's worry is greatly misplaced. Similarly, if one eats the poor modern diet, resides on a sofa, or drinks alcohol in copious amounts, one may worry more about these self induced risks than about deaths from climate change. Toxins as well as microbes from insufficient washing of one's hands or insufficient cooking of one's food cause orders of magnitude more deaths than does weather. Similarly, one should be more concerned with driving safely, not shooting one's self, practicing safe sex, and avoiding addiction to drugs before worrying about deaths from the atmosphere. In all, even if climate change were to increase weather deaths dramatically, those deaths would pale in comparison to the much more significant personally preventable causes of death.

CO$_2$ Increases Plant Growth

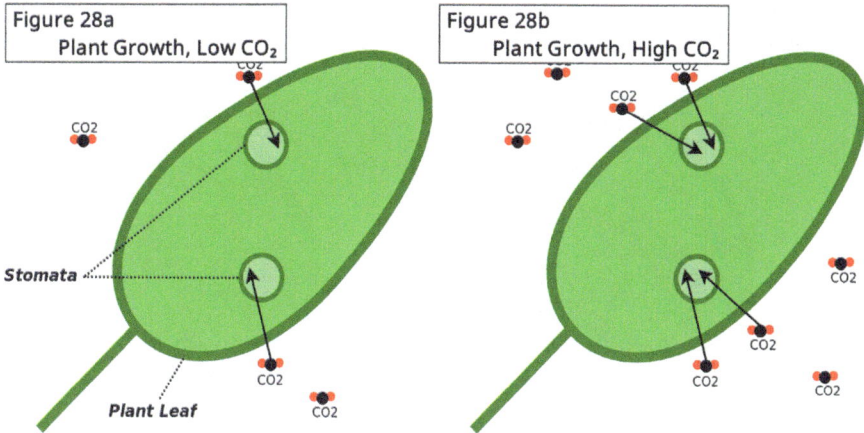

Figure 28a
Plant Growth, Low CO$_2$

Figure 28b
Plant Growth, High CO$_2$

Stomata

Plant Leaf

From early education, one may recall the process of photosynthesis. Plants use solar energy, water and carbon dioxide to photosynthesize the chemical energy of carbohydrates. The process of carbon dioxide uptake is illustrated in Figure 28. Plants take in carbon dioxide through small openings on the undersides of leaves, known as the stomata. When more carbon dioxide is available, carbon dioxide is more readily taken in through the stomata. Numerous studies confirm that when plants grow in controlled environments with increased carbon dioxide, photosynthesis increases and increased plant growth and crop yields result.[28.1]

Among the first scientists to advance the theory of global warming from increased carbon dioxide was the Swedish Nobel Laureate, Svante Arrhenius.[28.2] He also identified the so called fertilization effect of increased carbon dioxide. Arrhenius wrote: *"An increase of the carbon dioxide percentage to double its amount may hence be able to raise the intensity of vegetable life and the intensity of the inorganic chemical reactions threefold."*[28.3]

CO₂ & Drought Tolerance

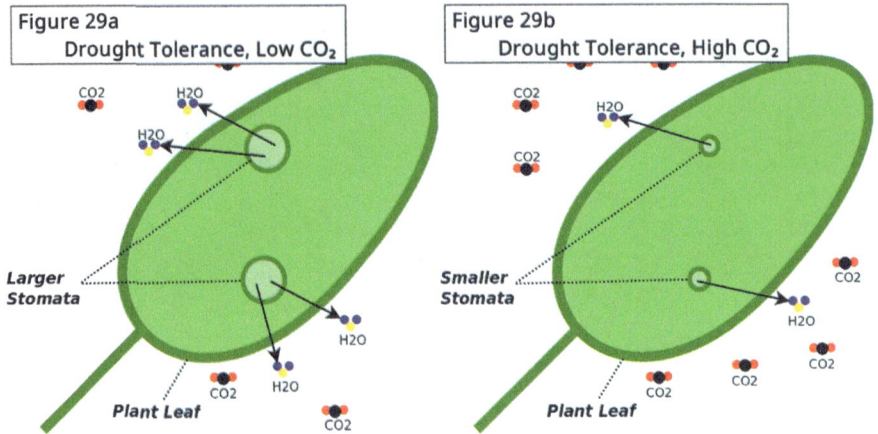

Figure 29a
    Drought Tolerance, Low CO₂

Figure 29b
    Drought Tolerance, High CO₂

When plant leaf stomata are open to take in carbon dioxide, they also allow water vapor to escape. Stomata may open or close, depending on the plant's need for carbon dioxide. When carbon dioxide concentration increases, the stomata may partially close but still receive carbon dioxide. Because of this partial closure, plants preserve water by limiting moisture loss. This is illustrated in Figure 29. To the extent this process prevails, increased carbon dioxide will tend to decrease water loss, and increase plant drought tolerance.[29.1]

Global Plant Growth Is Increasing

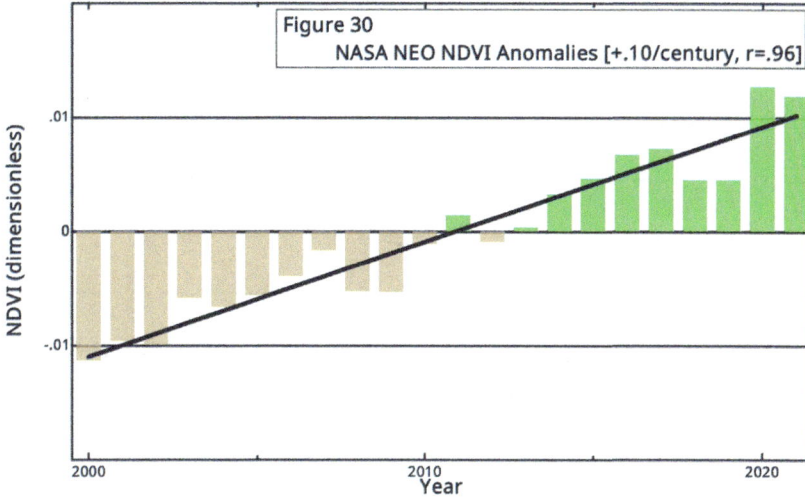

Figure 30
NASA NEO NDVI Anomalies [+.10/century, r=.96]

Satellite observations of a parameter called the Leaf Area Index indicate that plant growth around the world increased from 1999 through 2015.[30.1] A related and more recent measure from NASA is the Normalized Digital Vegetation Index (NDVI). The NDVI series appears in Figure 30.[30.2] There is a positive trend in this vegetation index. The trend of increased NDVI coincides with increased carbon dioxide and increased global temperature. This increase is consistent with the effects of increased photosynthesis and increased drought tolerance mentioned earlier.

Increased Global Phytoplankton

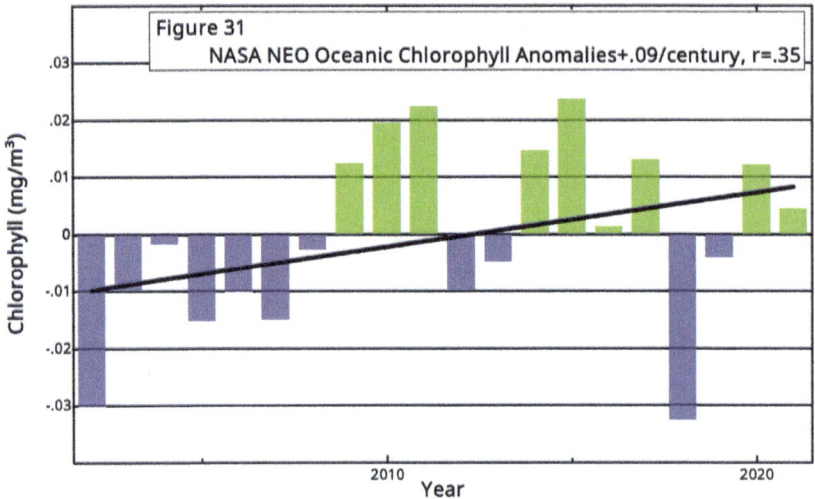

Figure 31
NASA NEO Oceanic Chlorophyll Anomalies+.09/century, r=.35

As with land-based vegetation, increasing levels of carbon dioxide may increase photosynthesis in ocean based single celled phytoplankton.[31.1] There is research indicating that phytoplankton increase has occurred in the Atlantic Ocean as well as in the Southern Ocean around Antarctica.[31.2,] [31.3] This is important because phytoplankton are at the base of the oceanic food web. Most ocean animal life depends directly or indirectly on phytoplankton. As noted earlier, the uptake of carbon dioxide from the atmosphere has increased. An increasing amount of carbon dioxide going into ocean life may limit the amount remaining in the atmosphere.

Figure 31 represents the global annual oceanic chlorophyll anomalies.[31.4] The trend of oceanic chlorophyll is positive, however, the correlation of this trend is much weaker than that of land based vegetation. This may be expected as land based vegetation is fixed in place by the roots of various plant life. Oceanic phytoplankton, on the other hand, drift with ocean currents. The phytoplankton may drift to locations which are more beneficial, or perhaps more detrimental than the original location. As a consequence, from year to year, phytoplankton may vary from factors other than the fertilization due to additional carbon dioxide.

Climate Models & Temperature

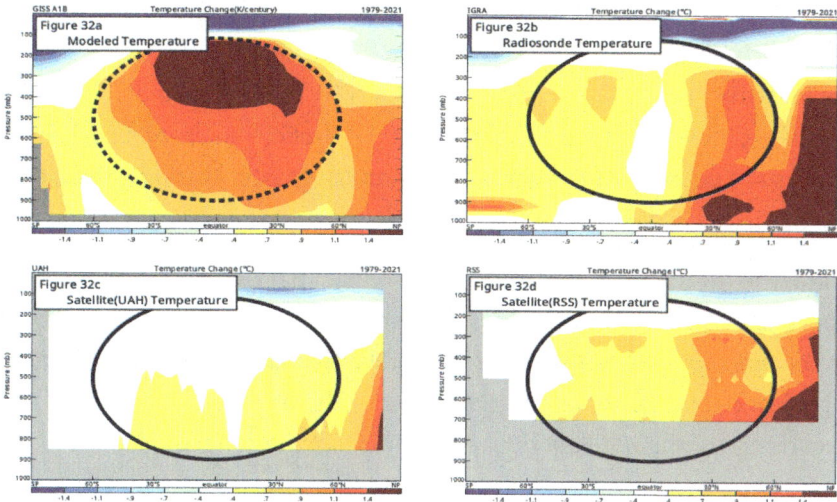

Figure 32a
Modeled Temperature

Figure 32b
Radiosonde Temperature

Figure 32c
Satellite(UAH) Temperature

Figure 32d
Satellite(RSS) Temperature

Early, simple estimates of global warming were of a motionless atmosphere at a single point. How much energy earth sends to space depends on the variation with height of temperature, greenhouse gasses, clouds, and other aerosols. In turn, these factors depend on the motions of the atmosphere. While the effect of increased carbon dioxide tends towards warming for most locations, motion of the atmosphere can determine where and how much warming might occur. Estimates turned to so called general circulation models to attempt to model what changes might occur with increased carbon dioxide.

The results of one such model compared with observed temperature trends are shown in Figure 32.[32.1], [32.2], [32.3], [32.4], [32.5], [32.6], [32.7] Figure 32a depicts temperature trends from a general circulation model. Figure 32b depicts temperature trends from weather balloon-based observations. Figure 32c depicts temperature trends from one satellite based analysis. Figure 32d depicts temperature trends from another satellite based analysis. There are three features of note.

First, the observations tend to match the model for the high layers at the top of each plot. These layers are at the stratosphere and exhibit a decreasing temperature trend. This is not a contradiction of global warming, but rather a confirmation of it. Because of the temperature profile of earth's atmosphere, additional carbon dioxide tends to cool the stratosphere.

Second, the observations tend to confirm the pronounced modeled increase of warming near the surface at the Arctic. In fact, while other factors are possible, the observations indicate greater warming than does the model.

Third, the model indicates a large area of maximal warming trend in the upper troposphere from about 60° South to 60° North. This area is known as the *hot spot*. The *hot spot* is indicated by the large oval on each plot. Significantly, the observations do not verify the *hot spot*. The lack of occurrence of the *hot spot* does not contradict the concept of global warming. But because the *hot spot* has so far not appeared, it indicates that the models have not accurately predicted the motions which would create the *hot spot*. Events such as clouds, storms, precipitation, droughts, floods, heat waves and cold waves are all determined by motions of the atmosphere. The failure of predicting the motions of the *hot spot* raises uncertainty about climate forecasts of all the other significant events which depend on motion.

Climate Models & Humidity

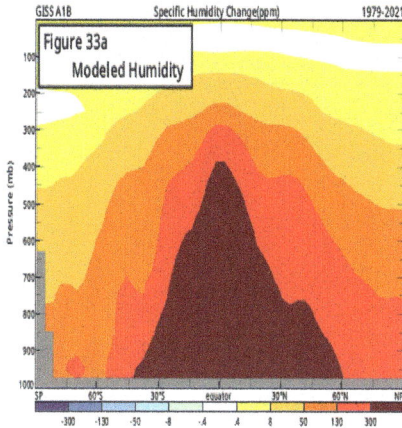

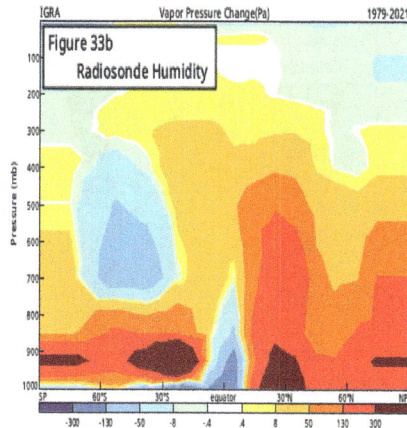

Figure 33a
Modeled Humidity

Figure 33b
Radiosonde Humidity

Another parameter climate models produce is humidity. Figure 33a represents the modeled trends of humidity.[33.1] Figure 33b represents the humidity trends from a related weather balloon measure.[33.2] As with any observational data, there are uncertainties with the balloon data. The modeled humidity trends exhibit three features. First, modeled humidity trends are greatest at the lowest levels and decrease with height. Second, modeled humidity trends are greatest at the tropics and decrease toward the poles. Third, modeled humidity trends are positive or close to zero at all levels and locations.

In contrast to the models, observations indicate large areas of decreasing humidity. These areas of decreasing humidity are delineated by the green and blue shaded areas of Figure 33b. These areas tend to be for the middle and upper troposphere. Part of the conceptual model of climate change is that global warming would, as modeled, increase absolute humidity throughout the troposphere leading to a positive water vapor feedback. Though partial, the areas of decreasing water vapor indicate increased outgoing infra-red radiation in these regions and some areas of negative feedback.

Climate Models & Wind Speed

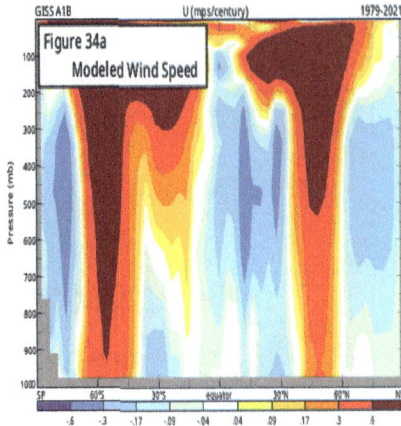

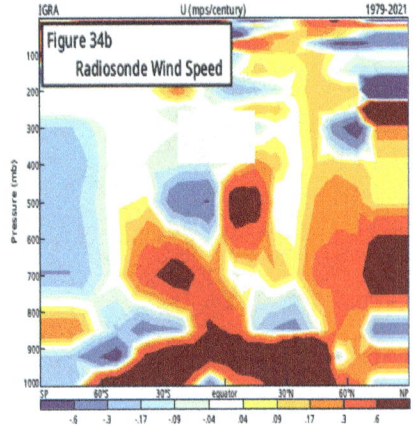

Figure 34a
Modeled Wind Speed

Figure 34b
Radiosonde Wind Speed

The trends of balloon-based wind speed measurements appear in Figure 34b in comparison to the modeled trends of wind speed in Figure 34a.[34.1], [34.2] The wind speeds presented are the so called zonal wind speeds, which refers to just the portion of the wind blowing eastward or westward. The modeled wind speeds indicate two large dark red areas representing trends of increased wind speed aloft from about thirty degrees to sixty degrees in each hemisphere. The model also indicates decreased wind speed trends in the lower atmosphere for both polar and equatorial regions. Weather and the averages of weather we call climate depend upon the motions of the atmosphere which are reflected by average wind speeds. Repeated again, there are uncertainties with all atmospheric observations. In this case, the balloon-based zonal wind speed trends are spatially incoherent. This may be due to the relative lack of spatial coverage by weather balloon observations.

US State Minimum Temperatures

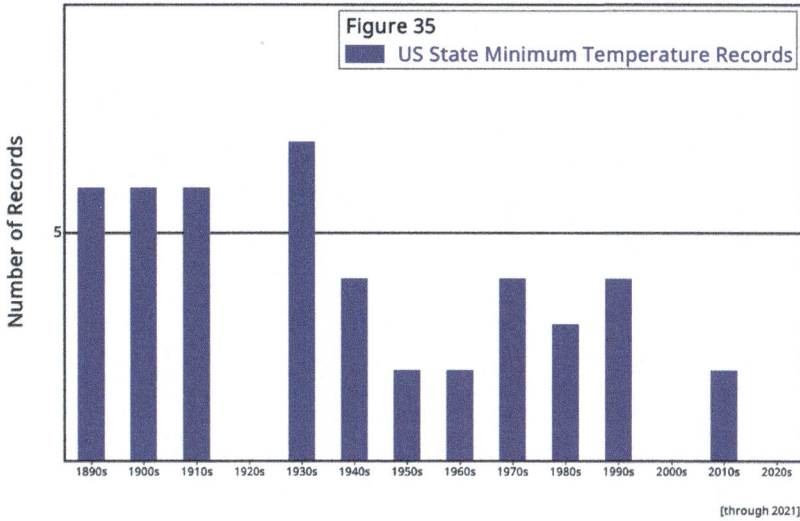

Figure 35
US State Minimum Temperature Records

Number of Records

1890s 1900s 1910s 1920s 1930s 1940s 1950s 1960s 1970s 1980s 1990s 2000s 2010s 2020s

[through 2021]

Agriculture has been significant to the history and development of the US. In turn, weather and climate are significant to agriculture. It's not surprising that notable farmer Thomas Jefferson promoted the idea of a weather observation network in 1776.[35.1] It took more than a century, but in 1890 the US Weather Bureau was formed and widespread standardized daily observations of weather began.[35.2]

Figure 35 is a graph of the occurrence of all time record low minimum temperatures observed in the US on a state by state basis by decade.[35.3] The record low minimum temperature for a state represents the lowest minimum temperature among the reporting stations for that state. The decade of the 1930s, despite its apt characterization as the Dust Bowl era, also experienced eight statewide all time low temperature records. The 1920s and the 2000s did not experience statewide all time record low temperatures. Fluctuations occur which create the conditions of extreme cold, but there is no apparent trend of the occurrence of statewide record low temperatures in the US associated with global warming.

US State Maximum Temperatures

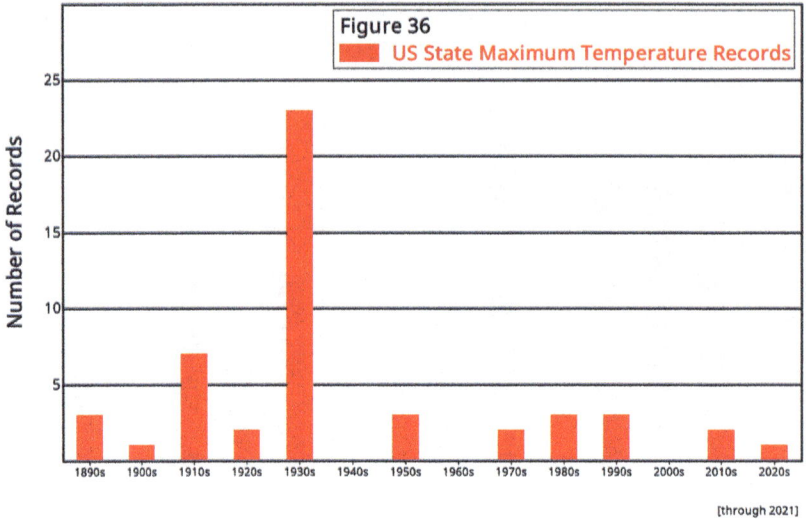

Figure 36 is a graph of the occurrence of all time record high maximum temperatures observed in the US on a state by state basis by decade.[36.1] The record high maximum temperature for a state represents the highest maximum temperature among the reporting stations for that state.

The 1930s experienced the greatest number of statewide record maximums. The 1940s, 1960s, and 2000s, did not experience statewide all time record high maximum temperatures. There is no apparent increase of the occurrence of statewide record high maximum temperatures in the US with global warming.

Fewer US Days Above 100°F

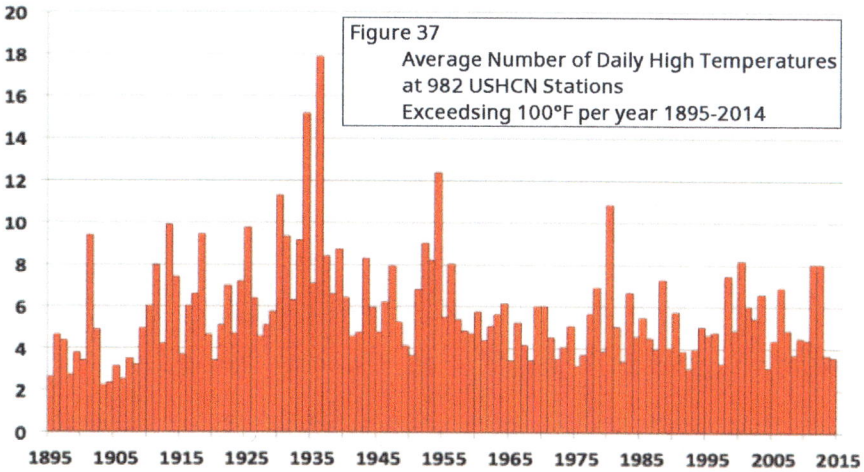

Figure 37
Average Number of Daily High Temperatures at 982 USHCN Stations Exceedsing 100°F per year 1895-2014

One may consider a broader measure of extreme heat by counting the number of days per year that US stations exceed a certain threshold of temperature. In 2016, Dr. John Christy of the University of Alabama at Hunstville provided such a measure. As part of his testimony before the US House Committee on Science, Space & Technology, Dr. Christy presented the figure above.[37.1] This figure represents the number of observations for all stations exceeding 100°F divided by the number of stations. The greatest number of hot days in the US occurred during the Dust Bowl era of the 1930s. The years with the greatest number of hot days were 1934 and 1936. There are uncertainties, but in this record, there is no apparent increase of hot days in the US associated with global warming.

US Hot Months

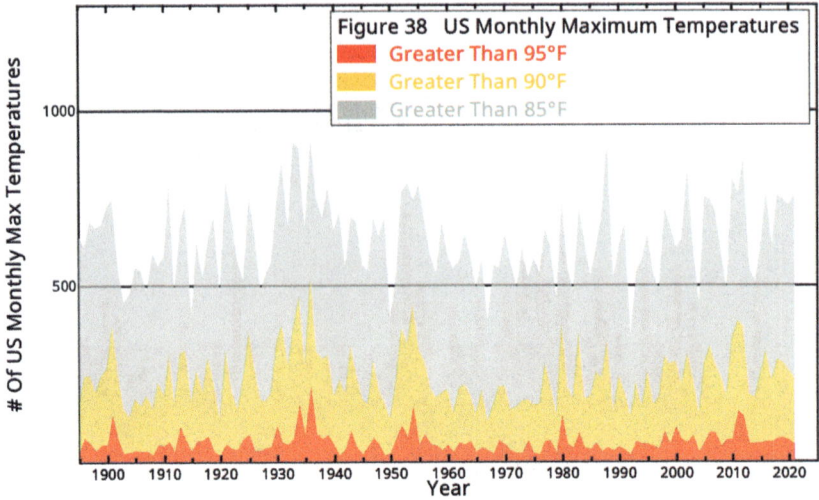

Figure 38 US Monthly Maximum Temperatures
- Greater Than 95°F
- Greater Than 90°F
- Greater Than 85°F

Instead of assessing the number of hot days, Figure 38 represents the number of hot months in the US.[38.1] This figure represents the number of monthly adjusted temperature observations for all stations exceeding thresholds. The greatest number of hot months in the US occurred in 1936 during the Dust Bowl. While there is much variation from year to year, the number of hot months in the US over recent decades does not appear to be unusual in the context of the long term record.

Variability May Decrease

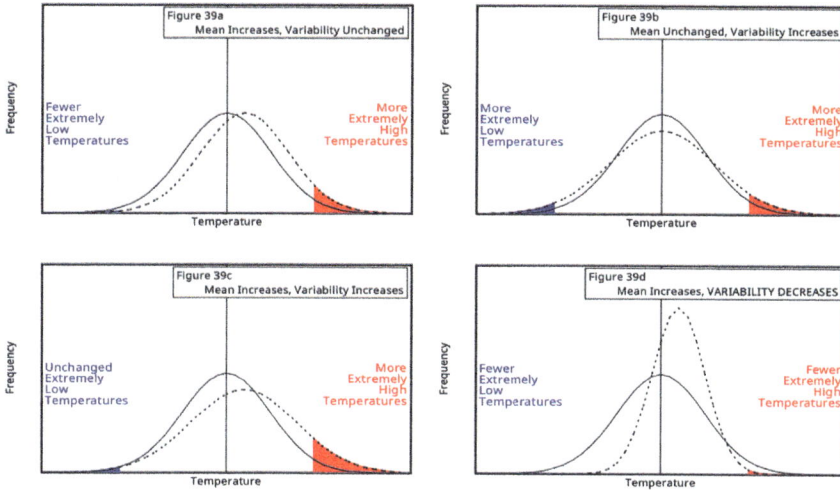

Figure 39a — Mean Increases, Variability Unchanged

Figure 39b — Mean Unchanged, Variability Increases

Figure 39c — Mean Increases, Variability Increases

Figure 39d — Mean Increases, VARIABILITY DECREASES

The preceding observations of extreme temperatures do not exhibit any apparent trend. At the same time, global warming has occurred and global average temperatures have increased. How might we explain this apparent paradox? It is possible that temperature variability may decrease as a consequence of mean global temperature increase. If this occurs, a reduction of extreme temperatures might be expected and not a paradox.

The distribution and frequency of parameters such as global temperature may be represented by the mean and the variance from that mean. In the Fifth Assessment Report, the IPCC considered changes to mean and variance.[39.1] Mean of temperature could increase, stay the same, or decrease. Variance of temperature could increase, stay the same, or decrease. The report depicted only cases of variance staying the same or increasing. The report did not depict the case of temperature variance decreasing. However, climate model results from decades earlier indicated reduced temperature variance with global warming.[39.2]

Figure 39 depicts how temperature extremes might result from changes to mean temperature and temperature variance.[39.3] Figure 39d depicts how temperature extremes might result from a combination of decreased temperature variance occurring with global warming. The solid line curves represent distributions of temperature occurrence at a starting state. The dotted line curves represent distributions of temperature occurrence with change. If variability remains the same or increases, more extreme high temperatures may result. However, as seen in Figure 39d, if mean temperature increases but temperature variance decreases, the result may be both fewer extreme low temperatures and fewer extreme high temperatures.

US Heating & Cooling Demand

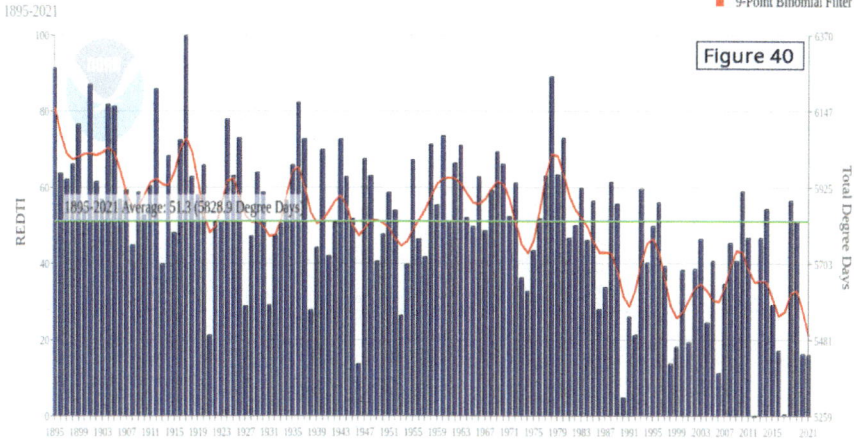

REDTI, Contiguous U.S.
1895-2021

Figure 40

Figure 40 is a plot from NOAA of the *Residential Energy Demand Temperature Index*[40.1] for the US.[40.2] This index represents hypothetical combined heating and air conditioning load. It is based on heating degree days or cooling degree days. This is calculated by the temperature difference from 65°F. The index is population weighted, meaning that part of the national index reflects more of areas where more people live and less of areas where fewer people live. The increase of average temperatures associated with global warming has meant increased air conditioning demand but also decreased heating demand. The lowest combined heating and cooling demand has occurred in the past three decades during the same time of global warming.

Fewer Strong US Tornadoes

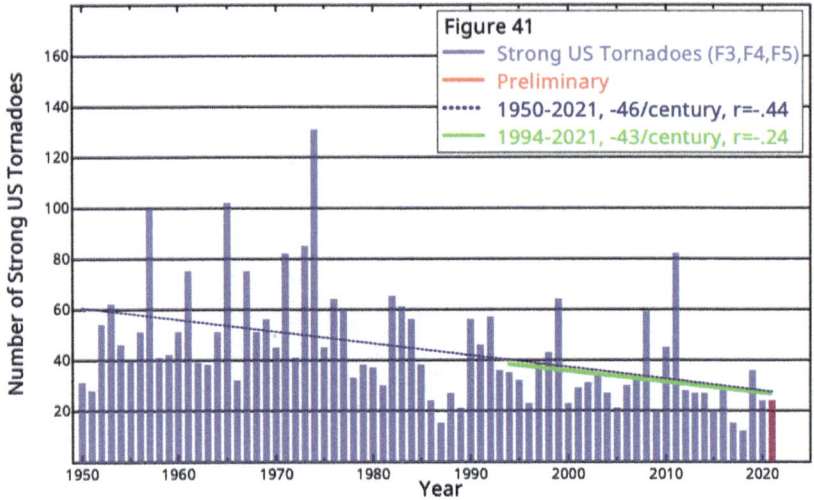

A conventional wisdom of climate change is that global warming will lead to increased events of extreme weather. This term is not concisely defined, but may include severe weather, and severe thunderstorms which often spawn tornadoes. Figure 41 is a plot of strong tornadoes in the US.[41.1] Strong tornadoes are defined by those rated as as category 3 or higher on the Enhanced Fujita Scale. The year of greatest number of strong tornadoes in the US was 1974. Counts from early in this record were assessed after the fact using estimates of damage to assign strength categories. By 1994, the NEXRAD Doppler radar array provided a more objective and complete record. The trends for the Doppler radar era are similar to the trends for the complete period of record. Because there cannot be fewer than zero strong tornadoes, such a decreasing trend cannot continue indefinitely. However, it appears that for the decades of the period of record, strong tornadoes in the US have occurred less frequently with global warming.

US Drought Has Not Increased

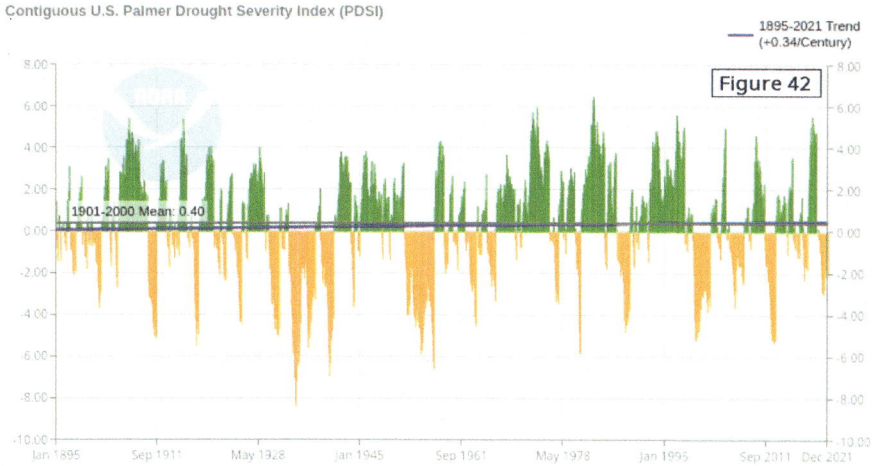

Contiguous U.S. Palmer Drought Severity Index (PDSI)

Figure 42

1895-2021 Trend (+0.34/Century)

1901-2000 Mean: 0.40

Should we expect a decrease of drought from global warming? There is evidence to support such a view. As we saw, the vegetation estimates over the last glacial cycle indicate that desert was prevalent during the cold ice age and forest was prevalent during the hot peak of the interglacial. There were, of course, confounding variables during these times. Motions of the atmosphere are thought to have changed as a consequence of the ice ages. Perhaps one of these changes was an increase of the intensity, extent, or duration of persistent subsidence. Dust in ice cores also indicates that the ice ages were much windier than present which increases evaporation. In any event, the vegetation record indicates much greater extent of extreme desert during the colder ice age and a much greater extent of closed forest during the warmth of the interglacial.

What do modern records indicate about drought? Figure 42 is a plot of the Palmer Drought Severity Index for the Contiguous US.[42.1] This index uses precipitation, temperature, and soil moisture data to estimate drought conditions lasting multiple months. The Dust Bowl era of the 1930s coincides with the greatest and most prolonged drought. There is no apparent increase of the frequency or intensity of drought in the US. Global estimates are based on sparse data of a briefer period. However, at least one study for 1982 through 2012 found no increase of drought in the global record.[42.2]

US Crop Moisture Stress

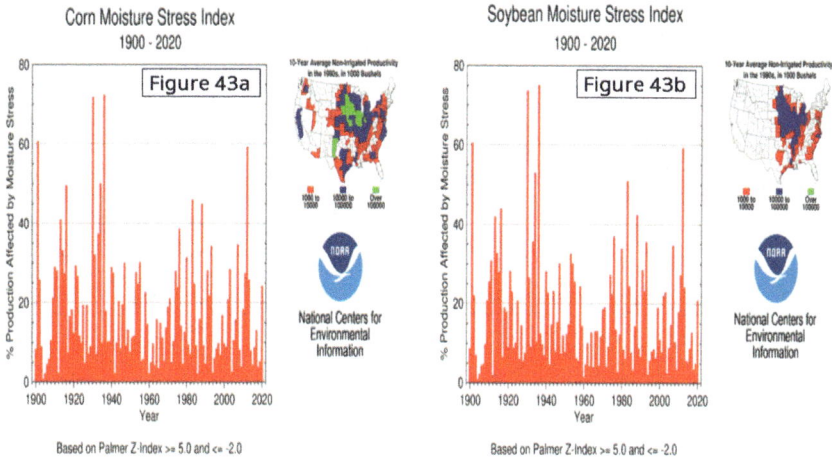

Figures 43a and 43b depict the US *Crop Moisture Stress Index* for the significant areas where corn and soybeans are grown.[43.1] This index is based on a drought measure which does not integrate past precipitation but considers only monthly data. As such, this index captures conditions which might last only one month. The crop growing areas for which these indices are calculated are depicted by the shaded areas on the small inset maps to the right of each plot. Consistent with the previous measures of drought and extreme heat, the years of the greatest crop moisture stress for the major growing areas were during the Dust Bowl of the 1930s. While there is a large amount of variation from year to year, there is no apparent increase of crop moisture stress in association with global warming. This data set was discontinued at the end of 2021.

Global Fires Have Decreased

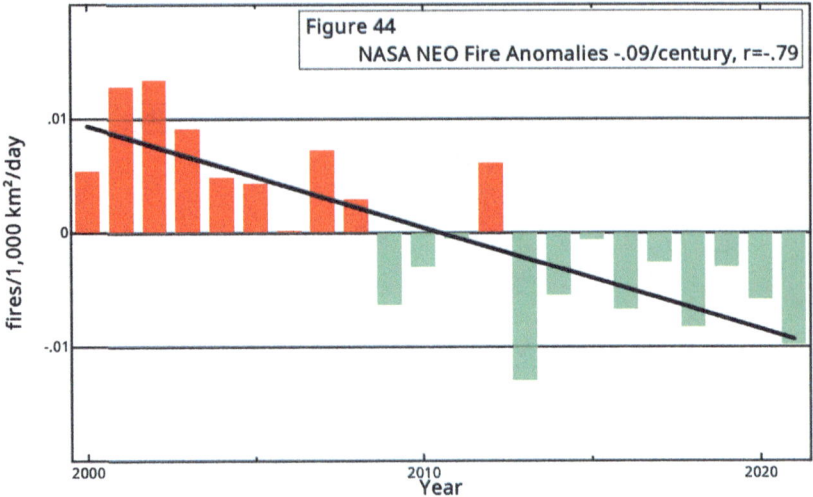

Figure 44
NASA NEO Fire Anomalies -.09/century, r=-.79

Another popular idea is that global warming will lead to increased wild fire. Fire frequency and intensity can be complex, but some of the factors are the frequency of ignition, the propensity to burn and spread, and the available fuel load. Ignition of wild fires can be both natural from lightning, and unnatural from humans. The propensity of fire to burn is largely a function of fire fuel load and fire fuel moisture. The propensity of fire to spread is largely a function of wind. Finally, the availability of fuel is determined largely by the number of fires in the years, decades, and centuries previous to a given event.

The US Forest Service was founded in 1905 including a mission of combating forest fires.[44.1] Fire suppression continues, in part motivated by the increasing number of people living in and around forest lands. However, there is growing recognition that fire has always been a natural part of forest ecology. So natural is fire to the forest that many species have evolved to only reproduce in the aftermath of fire.[44.2] Fire suppression may interfere with natural cycles of destruction and rebirth. In addition to interference with natural cycles, fire suppression has allowed fuel levels to increase. Suppressing fire in the short term increases the likelihood and intensity of fire in the longer term.

That climate change might change forest fires is perhaps based on the idea that drought might increase thus reducing the fuel moisture of forests and make fire more likely. Given the lack of evidence of drought increase in Figure 42, one might not expect any increase in forest fires.

Figure 44 is a plot of the global annual anomalies of fires per area per day.[44.3] This estimate from satellite observations covers the years 2001 through 2021. There are uncertainties and this twenty year record is still brief, but the negative trend of this data indicates a decrease in the number of fires globally.

Anecdote:
Bravado, Alcohol & Science

Another aspect of climate change assumed to occur with global warming is an increase in hurricane frequency or intensity. Hurricanes are episodic and relatively infrequent compared with mid latitude cyclones. The observations of hurricanes have changed over the past century. People have observed hurricanes from land throughout history, though high quality, high density systematic observations are relatively recent. Ship crews at sea logged storms, but for safety concerns, they actively avoided storms without measuring them. In 1943, aircraft made the first flights directly into a hurricane, actively pursuing rather than avoiding the storm. This leads us to the remarkable story of a remarkable man.

Joseph Duckworth was a pilot for the US Army Air Corps in 1927.[A1, A2] After his service, he flew commercial aircraft during which time he learned to fly by using instrument readings. During World War II, Duckworth was recalled to duty as a lieutenant colonel. Upon his return to service, Colonel Duckworth was shocked at the lack of knowledge about instrument flight. He trained pilots to use instrument flight, saving many lives in the process. In 1943, Colonel Duckworth was training British pilots on instrument techniques at Bryan Field in Texas. When the trainees learned that an approaching hurricane would cancel training and force evacuations, they began to tease Colonel Duckworth about the frailty of the training aircraft. Perhaps moved by national pride or by personal pride, Colonel Duckworth wagered the British pilots a cocktail that he could fly into the hurricane and safely return. On the historic first flight, Duckworth took with him navigator Lieutenant Colonel Ralph O'Hair who later recounted the story.[A3] The two were buffeted about but flew through the eye of the storm and returned! Upon returning, Duckworth made a second flight, this time with the weather officer, making the first aircraft based hurricane observations which are now regular and routine. Perhaps realizing he would meet resistance, Duckworth did not seek permission for this flight. One may imagine Duckworth's superior officers were peeved by his unauthorized flight, however, Colonel Duckworth received a medal rather than a reprimand for his adventure.

This story relates to how systematic active direct observations of hurricanes began. These observations have likely saved lives by more accurately tracking and forecasting hurricanes. But this story is also remarkable for the relationship between reason and emotion. Colonel Duckworth's advocacy of instrument flying was based on the reasoning that flying based on the objective measures from the instruments could improve safety as opposed to the more impulsive and emotional maneuvers of visual flying alone. At the same time, Duckworth's wager would appear to have been at least somewhat impulsive and emotional. He may have had a good understanding of the capabilities of the training aircraft and of how instrument flying could assist safety, but this first hurricane flight was not based on any evidence from a prior example. In this event, emotion and alcohol appear to have advanced science and understanding.

US Hurricanes Have Not Increased

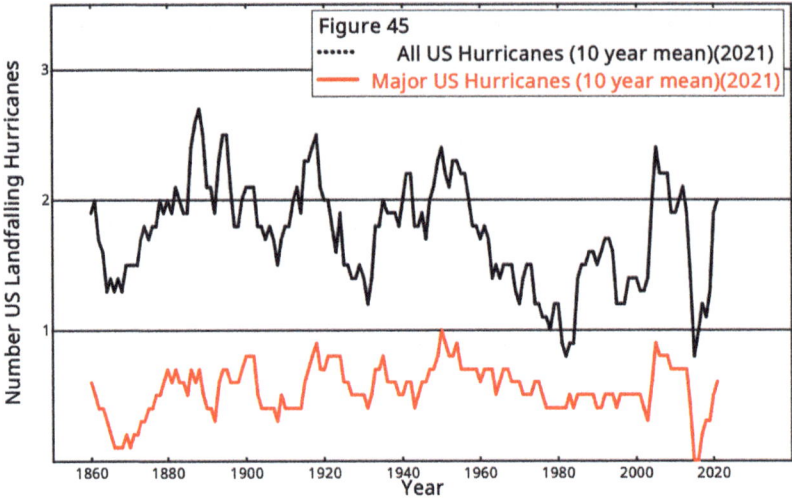

Figure 45
...... All US Hurricanes (10 year mean)(2021)
— Major US Hurricanes (10 year mean)(2021)

Direct observation by aircraft allows for the most accurate measurement of tropical cyclones but the long term global record is checkered by lack of observation and observation by different means. One long term record of note is of land falling hurricanes in the US. Because shipping has been so economically important, port cities have historically been the first to develop. And because weather was so important to shipping, port cities were places of regular weather observation. Figure 45 depicts the running ten year average number of hurricanes and major hurricanes making landfall on US coasts.[45.1] Such a measure necessarily excludes from consideration storms which dissipated or veered away from the US at sea. But the density of port cities enables an accurate and consistent measure of land falling storms. By this measure, peak hurricane frequency occurred before 1900. Major hurricane landfalls peaked in 1950. A period of nearly twelve years without a major hurricane landfall on US shores ended with Harvey in 2017. There is variation from decade to decade, but there is no apparent increase of land falling Atlantic US major or minor hurricanes associated with global warming.

Intensifying, Cooling Hurricanes

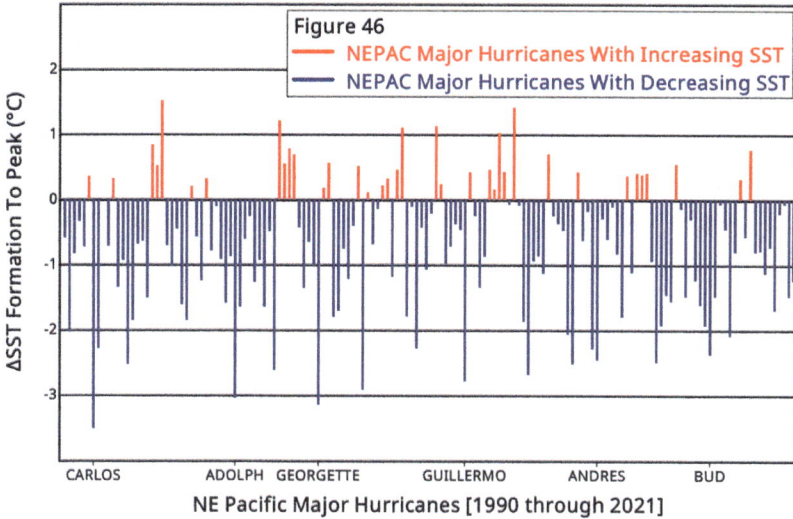

Figure 46
NEPAC Major Hurricanes With Increasing SST
NEPAC Major Hurricanes With Decreasing SST

NE Pacific Major Hurricanes [1990 through 2021]

There is speculation that hurricanes might increase in intensity with global warming because of increased sea surface temperatures. Figure 46 represents all the major hurricanes in the Northeast Pacific since 1990 and the change of sea surface temperatures over which the hurricanes intensified.[46.1], [46.2], [46.3], [46.4] The bars for each storm represent the change of sea surface temperature for each storm from formation of the storm to the peak intensity of the storm. Red bars indicate storms which intensified while traversing ocean waters which were warmer than those at formation. Blue bars indicate storms which intensified while traversing ocean waters which were cooler than those at formation. Most major hurricanes in the Northeast Pacific intensify even as the storms traverse a path of lower sea surface temperatures. For some storms, the decrease of sea surface temperatures exceeds 3°C! This effect is due to the spatial gradient of sea surface temperatures along the average paths of the storms in the Northeast Pacific. These contrary examples indicate other factors may be more significant than sea surface temperature to the intensification of tropical cyclones.

White Hat Bias

The Genesis Strategy:
Climate and Global Survival

Science as a Contact Sport: Inside the Battle to
Save the Earth's Climate

Storms of My Grandchildren: The Truth
About the Coming Climate Catastrophe
and Our Last Chance to Save Humanity

Figure 47 contains examples of titles of books and articles about climate change.[47.1, 47.2, 47.3] Who is not for "global survival"? Who opposes "saving the earth's climate"? Who opposes "saving humanity"?

In a 2010 paper, David Allison and Mark Cope defined *white hat bias* as distortion "in the service of what may be perceived as righteous ends."[47.4] If one believes one is pursuing righteous ends, by association one may believe one's self to be righteous. And if one believes one's self to be righteous, one may be more likely to dismiss contradictory evidence, which, by association, contradicts one's sense of righteousness. Do these titles about climate change evoke a sense of righteousness?

With respect to climate change, there may be accessories to consider other than white hats. When catastrophic climate scenarios are put forth, one may imagine them to be valid, and to imagine personal guilt at the hypothesized occurrence. Kahneman discusses how such a "selfish fear of regret" may be ultimately harmful by distorting risk and cost of avoidance.[47.5] Further, weather, climate, and climate change are complicated subjects. Even the very learned lack the time and energy to fully pursue many details. But one may feel obliged to have an opinion.

Deference to experts may occur out of fear of wearing the dunce cap, deference that may stifle further questioning and refined understanding. This may also be an example of a phenomenon observed by psychologist Solomon Asch in 1956. In a series of experiments, Asch noted the tendency of individuals to ignore their independent judgments and conform to a majority of peer subjects who falsely expressed an opposite, but distinctly incorrect judgment.[47.6]

Availability Bias

Figure 48a
Dust Bowl, 1930s

Figure 48b
Great Flood of 1862

Figure 48c
Galveston Hurricane, 1900

Figure 48d
Great Blizard of 1888

An important bias that may occur considering climate change is availability bias.[48.1] We have a tendency to emphasize recent events beyond historical ones. We have a tendency to emphasize visual and sensational portrayals of events beyond textual or numeric observations. We tend to consider only that which is more readily available and available to our senses. This is entirely understandable. Word of mouth accounts of long ago, perhaps passed on by legends, gave way to written accounts and measurements. In turn, written accounts were augmented by photographs and now nearly ubiquitous and instantaneous video. While the emotions surrounding modern events are understandable, we should remember records of past events. And we may reflect that many previous events occurred but were not recorded because the sparser populations of long ago lacked smart phones.

Figure 48 contains images of some notable US atmospheric events which are slowly fading from collective memory. These events occurred a century or more ago and are not necessarily proof of any aspect of the past climate or climate change. But how much more might we emphasize recent events because they are more available to us than are the photographs of historical events? And might this misplaced emphasis influence our tendency to confirm climate change theory?

Figure 48a is a photograph of the ravages from the Dust Bowl drought of the 1930s.[48.2] As discussed, many of the all time extreme temperature records of US stations are from the 1930s. There are temperature and precipitation data from this time, but the 1930s occurred before regular observations by weather balloons. As such, the circulation of the atmosphere during the time of the Dust Bowl is not well understood. The drought of the Dust Bowl coincided with greatly reduced precipitation, which is also thought to be the result of changes of circulation.

Figure 48b is a photograph of Sacremento, California in the aftermath of the Great Flood of 1862.[48.3] Weeks of heavy precipitation from December of 1861 through January of 1862 gave rise to catastrophic floods. Sediment records indicate that such events occur roughly every two centuries or so.[48.4] The cause of this unpredictable event is thought to be dynamic. That is to say, it was a fluctuation of atmospheric motion from ocean to land that likely caused such flooding. Global average temperature, which was lower than present, is not implicated. The sediment records indicate such an event will likely recur.

Figure 48c is a photograph of the destruction in the wake of The Galveston Hurricane of 1900.[48.5] Precise numbers of deaths are not known, but are thought to be around ten thousand. As a category four storm on the Saffir-Simpson scale, the hurricane was intense. Modern building construction, warnings and evacuations might have saved lives were they available at the time, but the storm remains the deadliest US hurricane.

Figure 48d depicts the Great Blizzard of 1888 which occurred March 11 through 14 of that year.[48.6] The blizzard shut down surface transportation and motivated construction of subway systems in Boston and New York.[48.7] The blizzard occurred as the result of an intense mid latitude cyclone known in the region as a Nor'Easter.

The events captured above by photograph should remind us that adverse weather and climate have always occurred. However, the amount and accessibility of emotionally compelling information for recent events continues to increase. Because of this increasing availability, we may overemphasize the significance of modern events compared to historical events.

Confirmation Bias

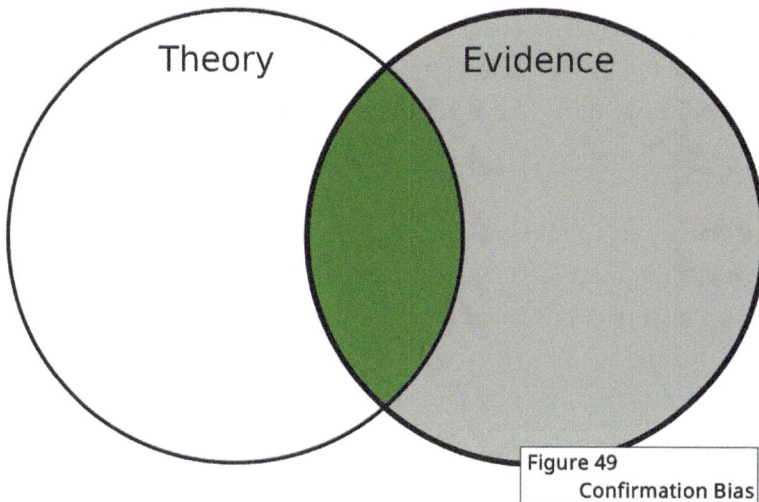

Figure 49
Confirmation Bias

Finally we come to confirmation bias, which is the tendency to consider evidence which confirms a theory and to exclude or minimize evidence which falsifies a theory. Though this tendency has been observed by many writers through history, the term was coined by British psychologist Peter Wason who conducted formal experiments exhibiting the behavior.[49.1] Wason systematically tested subjects by presenting them with a single series of numbers. He then asked subjects to determine the pattern of the series by having them present further examples. Subjects were far more likely to submit examples which matched a pattern they had in mind rather than to submit examples which would reject that pattern.

Those proposing a theory tend to emphasize observations of reality which support the theory. Advocates of alternative theories tend to emphasize evidence which supports such alternative theories. Emotional attraction to a theory may lead to the persistence of that theory and prevent true understanding. At the same time, however, when multiple theories openly compete, passion for those theories may lead to more energetic pursuit of observations. More complete observations and a willingness to test and accept them leads to a more complete understanding.

Concluding Remarks

That **Confirmation Bias** constitutes the final reason is by design. While the preconception that climate change is a worrisome problem can alter one's opinion of observations, it is not lost on the author that the contrary thesis, implicit in this book, is also susceptible to bias. Regarding differing views of objective data, Charles Darwin wrote the following in letter to a friend:

"About thirty years ago there was much talk that geologists ought only to observe and not theorise; and I well remember some one saying that at this rate a man might as well go into a gravel-pit and count the pebbles and describe the colours. How odd it is that anyone should not see that all observation must be for or against some view if it is to be of any service!"[CR1]

As with any written thesis, the reasons to worry less about climate change constitute a view, and views are subject to subjectivity. Still, the reasons above reflect common objective measures. I believe that accurate understanding of climate change requires continued ongoing consideration of all these objective measures.

An important aspect of climate change to keep in mind is that climate change observations constitute relatively weak evidence. In medicine, the highest standard of evidence is so called *meta-analysis of replicated, randomized, blinded, controlled experiments*. With respect to climate change, other factors are not constrained, so the experiment is not controlled. With respect to climate change, the experiment is singular, and is without randomization or replication, and so, meta-analysis is not possible. Further, since climate change predictions are of the future, even a single instance of the experiments is not complete. And because climate models and observations are public, studies of climate change are not blinded. While these factors relegate testing of theories of climate change to the relatively weaker category of *observational studies*, they are all we have.

All theses are subject to continued observation. In order to continue such observations, I intend to regularly update this book with the latest available analysis. The reader is invited to continue to assess these objective measures against the popular understanding of climate change.

Notes

[1.1] James Hansen, "Testimony to the United States Senate Committee on Energy and Natural Resources," June 23, 1988, https://www.sealevel.info/1988_Hansen_Senate_Testimony.pdf.

[1.2] James Hansen, I. Fung, A. Lacis, D. Rind, S. Lebedeff, R. Ruedy, G. Russell, and P. Stone, "Global climate changes as forecast by Goddard Institute for Space Studies three-dimensional model," *J. Geophys. Res.* 93,D8 (August 20, 1988): 9341-9364, doi:10.1029/JD093iD08p09341. Data accessed October 24, 2019, http://www.realclimate.org/data/scen_ABC_temp.data. Lines indicate trends from ordinary least squares regression.

[1.3] NOAA NCEI, "Global Land and Ocean Temperature Anomalies," accessed June 24, 2021, https://www.ncdc.noaa.gov/cag/global/time-series/globe/land_ocean/12/12/1880-2021/data.csv. Lines indicate trends from ordinary least squares regression.

[2.1] NOAA NCEI, "Global Land and Ocean Temperature Anomalies," accessed June 24, 2021, https://www.ncdc.noaa.gov/cag/global/time-series/globe/land_ocean/12/12/1880-2021/data.csv. Plotted values indicate trailing forty year trends from ordinary least squares regression.

[3.1] NOAA NCEI, "Global Land and Ocean Temperature Anomalies," accessed June 24, 2021, https://www.ncdc.noaa.gov/cag/global/time-series/globe/land_ocean/12/12/1880-2021/data.csv. Plotted values indicate trailing forty year trends from ordinary least squares regression.

[3.2] IPCC, *Climate Change 2007: Synthesis Report. Contribution of Working Groups I, II and III to the Fourth Assessment Report of the Intergovernmental Panel on Climate Change [Core Writing Team, Pachauri, R.K and Reisinger, A. (eds.)]* (Geneva, Switzerland: IPCC, 2007): 45, https://www.ipcc.ch/site/assets/uploads/2018/02/ar4_syr_full_report.pdf .

See Table 3.1.

[4.1] James Hansen, I. Fung, A. Lacis, D. Rind, S. Lebedeff, R. Ruedy, G. Russell, and P. Stone, "Global climate changes as forecast by Goddard Institute for Space Studies three-dimensional model," *J. Geophys. Res.* 93,D8 (August 20, 1988): 9341-9364, doi:10.1029/JD093iD08p09341. Modeled radiative forcing scenario data accessed October 24, 2019, http://www.realclimate.org/data/H88_scenarios_eff.dat.

[4.2] James H. Butler and Stephen A. Montzka, "NOAA Annual Greenhouse Gas Index(AGGI)," Spring 2022, https://www.esrl.noaa.gov/gmd/aggi/aggi.html. Data are with respect to change since 1988. Trends are ordinary least squares regression for 1988 through 2018.

[5.1] James H. Butler and Stephen A. Montzka, "NOAA Annual Greenhouse Gas Index(AGGI)," Spring 2022, https://www.esrl.noaa.gov/gmd/aggi/aggi.html. Plotted values indicate trailing five year trends from ordinary least squares regression.

[6.1] Crippa, M., Guizzardi, D., Solazzo, E., Muntean, M., Schaaf, E., Monforti-Ferrario, F., Banja, M., Olivier, J.G.J., Grassi, G., Rossi, S., Vignati, E., "GHG emissions of all world countries - 2021 Report," EUR 30831 EN, Publications Office of the European Union, Luxembourg, 2021, ISBN 978-92-76-41547-3, doi:10.2760/173513, JRC126363. Country emissions data accessed January 10, 2022, https://edgar.jrc.ec.europa.eu/booklet/EDGARv6.0_FT2020_fossil_CO2_GHG_booklet2021.xls.

[7.1] Crippa, M., Guizzardi, D., Solazzo, E., Muntean, M., Schaaf, E., Monforti-Ferrario, F., Banja, M., Olivier, J.G.J., Grassi, G., Rossi, S., Vignati, E., "GHG emissions of all world countries - 2021 Report," EUR 30831 EN, Publications Office of the European Union, Luxembourg, 2021, ISBN 978-92-76-41547-3, doi:10.2760/173513, JRC126363. Country emissions data accessed January 10, 2022, https://edgar.jrc.ec.europa.eu/booklet/EDGARv6.0_FT2020_fossil_CO2_GHG_booklet2021.xls.

[8.1] Crippa, M., Guizzardi, D., Solazzo, E., Muntean, M., Schaaf, E., Monforti-

Ferrario, F., Banja, M., Olivier, J.G.J., Grassi, G., Rossi, S., Vignati, E., "GHG emissions of all world countries - 2021 Report," EUR 30831 EN, Publications Office of the European Union, Luxembourg, 2021, ISBN 978-92-76-41547-3, doi:10.2760/173513, JRC126363.
Country emissions data accessed January 10, 2022, https://edgar.jrc.ec.europa.eu/booklet/EDGARv6.0_FT2020_fossil_CO2_GHG_booklet2021.xls.

[9.1] World Bank, "Fertility rate, total (births per woman)," accessed April 14, 2022, https://data.worldbank.org/indicator/SP.DYN.TFRT.IN.

[9.2] George Friedman, *The next 100 years: a forecast for the 21st century* (New York: Anchor Books, 2010) 50-56.

[10.1] CIA, "The World Factbook, Total Fertility Rate" (Washington, DC: Central Intelligence Agency), accessed April 11, 2022, https://www.cia.gov/the-world-factbook/field/total-fertility-rate/country-comparison.

[11.1] World Bank, "Fertility rate, total (births per woman)," accessed April 14, 2022, https://data.worldbank.org/indicator/SP.DYN.TFRT.IN.

[11.2] UN, Department of Economic and Social Affairs, Population Division, "Probabilistic Population Projections Rev. 1 based on the World Population Prospects 2019 Rev. 1," http://population.un.org/wpp/. Median fertility estimates accessed June 24, 2021, https://population.un.org/wpp/Download/Files/2_Indicators%20(Probabilistic%20Projections)/UN_PPP2019_Input_TFR.xlsx.

[12.1] Crippa, M., Guizzardi, D., Solazzo, E., Muntean, M., Schaaf, E., Monforti-Ferrario, F., Banja, M., Olivier, J.G.J., Grassi, G., Rossi, S., Vignati, E., "GHG emissions of all world countries - 2021 Report," EUR 30831 EN, Publications Office of the European Union, Luxembourg, 2021, ISBN 978-92-76-41547-3, doi:10.2760/173513, JRC126363.
Country emissions data accessed January 10, 2022, https://edgar.jrc.ec.europa.eu/booklet/EDGARv6.0_FT2020_fossil_CO2_GHG_booklet2021.xls.

[12.2] CIA, "The World Factbook, Total Fertility Rate" (Washington, DC:

Central Intelligence Agency), accessed April 11, 2022, https://www.cia.gov/the-world-factbook/field/total-fertility-rate/country-comparison.

[13.1] Friedlingstein, P., Jones, M. W., O'Sullivan, M., Andrew, R. M., Bakker, D. C. E., Hauck, J., Le Quéré, C., Peters, G. P., Peters, W., Pongratz, J., Sitch, S., Canadell, J. G., Ciais, P., Jackson, R. B., Alin, S. R., Anthoni, P., Bates, N. R., Becker, M., Bellouin, N., Bopp, L., Chau, T. T. T., Chevallier, F., Chini, L. P., Cronin, M., Currie, K. I., Decharme, B., Djeutchouang, L., Dou, X., Evans, W., Feely, R. A., Feng, L., Gasser, T., Gilfillan, D., Gkritzalis, T., Grassi, G., Gregor, L., Gruber, N., Gürses, Ö., Harris, I., Houghton, R. A., Hurtt, G. C., Iida, Y., Ilyina, T., Luijkx, I. T., Jain, A. K., Jones, S. D., Kato, E., Kennedy, D., Klein Goldewijk, K., Knauer, J., Korsbakken, J. I., Körtzinger, A., Landschützer, P., Lauvset, S. K., Lefèvre, N., Lienert, S., Liu, J., Marland, G., McGuire, P. C., Melton, J. R., Munro, D. R., Nabel, J. E. M. S., Nakaoka, S.-I., Niwa, Y., Ono, T., Pierrot, D., Poulter, B., Rehder, G., Resplandy, L., Robertson, E., Rödenbeck, C., Rosan, T. M., Schwinger, J., Schwingshackl, C., Séférian, R., Sutton, A. J., Sweeney, C., Tanhua, T., Tans, P. P., Tian, H., Tilbrook, B., Tubiello, F., van der Werf, G., Vuichard, N., Wada, C., Wanninkhof, R., Watson, A., Willis, D., Wiltshire, A. J., Yuan, W., Yue, C., Yue, X., Zaehle, S., and Zeng, J.: "Global Carbon Budget 2021," *Earth System Science Data* Discuss. [preprint], https://doi.org/10.5194/essd-2021-386, in review, 2021.

[13.2] Global Carbon Project, "Supplemental data of Global Carbon Budget 2021 (Version 1.0) [Data set]," Global Carbon Project, data accessed January 11, 2022, doi:10.18160/gcp-2021.

[14.1] CDIAC, "Historical Isotopic Temperature Record from the Vostok Ice Core," *Carbon Dioxide Information Analysis Center* (2000), https://cdiac.ess-dive.lbl.gov/.
Vostok ice core proxy temperature data accessed November 30, 2019: http://cdiac.ess-dive.lbl.gov/ftp/trends/temp/vostok/vostok.1999.temp.dat.

[14.2] J. Jouzel, C. Lorius, J.R. Petit, C. Genthon, N.I. Barkov, V.M. Kotlyakov, and V.M. Petrov, "Vostok ice core: a continuous isotope temperature record over the last climatic cycle (160,000 years)," *Nature* 329 (October 1, 1987): 403-408, doi:10.1038/329403a0.

14.3 J. Jouzel, N.I. Barkov, J.M. Barnola, M. Bender, J. Chappellaz, C. Genthon, V.M. Kotlyakov, V. Lipenkov, C. Lorius, J.R. Petit, D. Raynaud, G. Raisbeck, C. Ritz, T. Sowers, M. Stievenard, F. Yiou, and P. Yiou. "Extending the Vostok ice-core record of palaeoclimate to the penultimate glacial period," *Nature* 364 (March 22, 1993): 407-412, doi:10.1038/364407a0.

14.4 J. Jouzel, J., C. Waelbroeck, B. Malaize, M. Bender, J.R. Petit, M. Stievenard, N.I. Barkov, J.M. Barnola, T. King, V.M. Kotlyakov, V. Lipenkov, C. Lorius, D. Raynaud, C. Ritz, and T. Sowers. "Climatic interpretation of the recently extended Vostok ice records," *Climate Dynamics* (12, June, 1996): 513-521, doi:10.1007/BF00207935.

14.5 J.R. Petit, J. Jouzel, D. Raynaud, N.I. Barkov, J.-M. Barnola, I. Basile, M. Bender, J. Chappellaz, M. Davis, G. Delayque, M. Delmotte, V.M. Kotlyakov, M. Legrand, V.Y. Lipenkov, C. Lorius, L. Pepin, C. Ritz, E. Saltzman, and M. Stievenard. "Climate and atmospheric history of the past 420,000 years from the Vostok ice core, Antarctica," *Nature* 399 (June 3, 1999): 429-436, doi:10.1038/20859.

15.1 CDIAC, "Historical Isotopic Temperature Record from the Vostok Ice Core," *Carbon Dioxide Information Analysis Center* (2000), https://cdiac.ess-dive.lbl.gov/.
Vostok ice core proxy temperature data accessed November 30, 2019: http://cdiac.ess-dive.lbl.gov/ftp/trends/temp/vostok/vostok.1999.temp.dat.

15.2 J. Jouzel, C. Lorius, J.R. Petit, C. Genthon, N.I. Barkov, V.M. Kotlyakov, and V.M. Petrov, "Vostok ice core: a continuous isotope temperature record over the last climatic cycle (160,000 years)," *Nature* 329 (October 1, 1987): 403-408,
doi:10.1038/329403a0.

15.3 J. Jouzel, N.I. Barkov, J.M. Barnola, M. Bender, J. Chappellaz, C. Genthon, V.M. Kotlyakov, V. Lipenkov, C. Lorius, J.R. Petit, D. Raynaud, G. Raisbeck, C. Ritz, T. Sowers, M. Stievenard, F. Yiou, and P. Yiou. "Extending the Vostok ice-core record of palaeoclimate to the penultimate glacial period," *Nature* 364 (March 22, 1993): 407-412, doi:10.1038/364407a0.

15.4 J. Jouzel, J., C. Waelbroeck, B. Malaize, M. Bender, J.R. Petit, M. Stievenard, N.I. Barkov, J.M. Barnola, T. King, V.M. Kotlyakov, V. Lipenkov, C. Lorius, D. Raynaud, C. Ritz, and T. Sowers. "Climatic interpretation of the recently extended Vostok ice records," *Climate Dynamics* 12 (June, 1996): 513-521, doi:10.1007/BF00207935.

15.5 J.R. Petit, J. Jouzel, D. Raynaud, N.I. Barkov, J.-M. Barnola, I. Basile, M. Bender, J. Chappellaz, M. Davis, G. Delayque, M. Delmotte, V.M. Kotlyakov, M. Legrand, V.Y. Lipenkov, C. Lorius, L. Pepin, C. Ritz, E. Saltzman, and M. Stievenard. "Climate and atmospheric history of the past 420,000 years from the Vostok ice core, Antarctica," *Nature* 399 (June 3, 1999): 429-436, doi:10.1038/20859.

16.1 Richard B. Alley, "GISP2 Ice Core Temperature and Accumulation Data," *IGBP PAGES/World Data Center for Paleoclimatology, Data Contribution Series #2004-013* (Boulder CO, USA: NOAA/NGDC Paleoclimatology Program, 2004), accessed March 1, 2022, https://www.ncei.noaa.gov/pub/data/paleo/icecore/greenland/summit/gisp2/isotopes/gisp2_temp_accum_alley2000.txt.

16.2 Richard Alley, "The Younger Dryas cold interval as viewed from central Greenland," *Quaternary Science Reviews* 19 (January 1, 2000): 213-226, doi:10.1016/S0277-3791(99)00062-1.

16.3 Jay Bennett, "The Time a P-38 Was Pulled From the Ice and Restored to Flying Condition," *PopularMechanics.com*, January 12, 2016, https://www.popularmechanics.com/flight/a18943/glacier-girl-p-38-fighter.

16.4 Joris Nieuwint, "15 Unbelievable Pictures From P-38 Glacier Girl You've Never Seen Before," *WarHistoryOnline.com*, February 10, 2016, https://www.warhistoryonline.com/military-vehicle-news/15-p-38-glacier-girl-pictures.html.

16.5 Tom Metcalfe, "'Lost Squadron' WWII Warplane Discovered Deep Beneath a Greenland Glacier," *LiveScience.com* August 25, 2018, https://www.livescience.com/63423-lost-squadron-unearthed-greenland-glacier.html.

[17.1] Early migrations mercator.svg. (2021, March 6). Wikimedia Commons, the free media repository. Retrieved April 1, 2021 from https://commons.wikimedia.org/wiki/File:Early_migrations_mercator.svg.

[18.1] J.M. Adams, "Global land environments since the last interglacial," *Oak Ridge National Laboratory* TN, USA, accessed October 24, 2019, https://www.esd.ornl.gov/projects/gen/nerc.html.
Images adapted from: https://www.esd.ornl.gov/projects/gen/lastgla.gif, https://www.esd.ornl.gov/projects/gen/earlyho.gif, https://www.esd.ornl.gov/projects/gen/midhol.gif, and https://www.esd.ornl.gov/projects/gen/pres-pot.gif.

[18.2] J.M. Adams and H. Faure (QEN members), "Review and Atlas of Palaeovegetation: Preliminary land ecosystem maps of the world since the Last Glacial Maximum," *Oak Ridge National Laboratory* TN, USA, accessed October 24, 2019, https://www.esd.ornl.gov/projects/gen/adams1.html.

[19.1] J. Laskar, P. Robutel, F. Joutel, M. Gastineau, A. C. M. Correia and B. Levrard, "A long term numerical solution for the insolation quantities of the Earth," *Astronomy and Astrophysics* 428, 1 (December 11, 2004): 261-285, doi:10.1051/0004-6361:20041335.

[19.2] Colorado State University, Denning Research Group, "Milankovitch Orbital Data Viewer," accessed November 30, 2019, https://biocycle.atmos.colostate.edu/shiny/Milankovitch/.

[19.3] Milutin Milanković, *Mathematische Klimalehre und astronomische Theorie der Klimaschwankungen, Volume 1, Part 1 of Handbuch der Klimatologie in fünf Bänden* (Berlin: Gebrüder Borntraeger, 1930).

[20.1] Hersbach, Hans & Bell, Bill & Berrisford, Paul & Hirahara, Shoji & Horányi, András & Muñoz Sabater, Joaquín & Nicolas, Julien & Peubey, Carole & Radu, Raluca & Schepers, Dinand & Simmons, Adrian & Soci, Cornel & Abdalla, Saleh & Abellan, Xavier & Balsamo, Gianpaolo & Bechtold, Peter & Biavati, Gionata & Bidlot, Jean & Bonavita, Massimo & Thépaut, J.-N. (2020). "The ERA5 global reanalysis." *Quarterly Journal of*

the *Royal Meteorological Society.* https://doi.org/10.1002/qj.3803.
Copernicus Climate Change Service (C3S) (2017): ERA5: Fifth generation of ECMWF atmospheric reanalyses of the global climate, *Copernicus Climate Change Service Climate Data Store (CDS)*, accessed January 6, 2022, https://cds.climate.copernicus.eu/cdsapp#!/home. Contains modified Copernicus Climate Change Service Information 2019.
Vectors of ERA5 surface wind anomalies are of ordinary least squares regression of U,V component winds for 1979 through 2021. The size of the vectors is proportionate to the relative magnitude of wind speed anomaly.

[21.1] Earth System Research Laboratory, (Global Radiation Group, Global Monitoring Division), "Atmospheric Transmission of Direct Solar Radiation at Mauna Loa, Hawaii," *National Oceanographic and Atmospheric Administration*, accessed January, 2022, https://www.esrl.noaa.gov/gmd/grad/mloapt.html.
Image adapted from:https://www.esrl.noaa.gov/gmd/webdata/grad/mloapt/mauna_loa_transmission.png.

[22.1] Kevin E. Trenberth, John T. Fasullo, and Jeffrey Kiehl, "Earth's global energy budget," *Bull. Amer. Meteorol. Soc.* 90, 3 (March 1, 2009): 314, doi:10.1175/2008BAMS2634.1.

[22.2] NASA, (Earth Observing System, Langley Research Center), "Clouds and the Earth's Radiant Energy System (CERES)," accessed January, 2022, https://ceres-tool.larc.nasa.gov/ord-tool/jsp/EBAFTOA41Selection.jsp.

[23.1] NOAA, (NESDIS Center for Satellite Applications and Research, Laboratory for Satellite Altimetry), "Global mean sea level from TOPEX/Poseidon, Jason-1, and Jason-2," accessed January 2022, https://www.star.nesdis.noaa.gov/sod/lsa/SeaLevelRise/.
Image adapted from:https://www.star.nesdis.noaa.gov/sod/lsa/SeaLevelRise/slr/slr_sla_g bl_free_txj1j2_90_400.png.

[23.2] Marc F. P. Bierkens and Yoshihide Wada, "Non-renewable groundwater use and groundwater depletion: a review," *Environ. Res. Lett.* 14, 6 (May 29, 2019), doi:10.1088/1748-9326/ab1a5f.

[23.3] Yoshihide Wada, L. P. H. van Beek, C. M. van Kempen, J. W. T. M. Reckman, S. Vasak, and M. F. P. Bierkens, "Global depletion of groundwater resources," *Geophys. Res. Lett.* 37, L20402, (2010), doi:10.1029/2010GL044571.

[23.4] Leonard F. Konikow, "Contribution of global groundwater depletion since 1900 to sea-level rise," *Geophys. Res. Lett.* 38, 17 (September 2, 2011), doi:10.1029/2011GL048604.

[23.5] Yoshihide Wada, Ludovicus P. H. van Beek, Weiland Sperna, C. Frederiek, Benjamin F. Chao, Yun-Hao Wu, Marc F. P. Bierkens, "Past and future contribution of global groundwater depletion to sea-level rise," *Geophys. Res. Lett.* 39, 9 (2012), doi:10.1029/2012GL051230.

[23.6] Yoshihide Wada, and Marc F P Bierkens, "Sustainability of global water use: past reconstruction and future projections," *Environ. Res. Lett.* 10, 9 (October 1, 2014), doi:10.1088/1748-9326/9/10/104003.

[23.7] S. Yoshikawa, J. Cho, H. G. Yamada, N. Hanasaki, and S. Kanae, "An assessment of global net irrigation water requirements from various water supply sources to sustain irrigation: rivers and reservoirs (1960–2050)," *Hydrol. Earth Syst. Sci.* 18, (2014): 4289–4310, doi:10.5194/hess-18-4289-2014.

[24.1] Centers for Disease Control and Prevention, (National Center for Health Statistics, National Vital Statistics System), "Deaths From Each Cause by Month, Race, and Sex: United States, 1999-2006," https://www.cdc.gov/nchs/nvss/mortality/gmwkiv.htm

[24.2] Matthias Wacker and Michael F Holick, "Sunlight and Vitamin D: A global perspective for health," *Dermato-endocrinology* 5, 1 (January 1, 2013): 51-108, doi:10.4161/derm.24494.

[24.3] Graham Holliman, Donna Lowe, Howard Cohen, Sarah Felton, and Ken Raj, "Ultraviolet Radiation-Induced Production of Nitric Oxide:A multi-cell and multi-donor analysis," *Sci Rep.* 7, 1 (September 11, 2017): 11105-11105, doi:10.1038/s41598-017-11567-5.

[24.4] M Nathaniel Mead, "Benefits of sunlight: a bright spot for human health," *Environmental Health Perspectives* 116, 4 (April 2008): A160-A167, doi:10.1289/ehp.116-a160.

25.1 Jeffrey Berko, Deborah D. Ingram, Shubhayu Saha, and Jennifer D. Parker, (CDC, NCHS), "Deaths Attributed to Heat, Cold, and Other Weather Events in the United States, 2006–2010," *National Health Statistics Reports* 76 (July 30, 2014): 5, https://www.cdc.gov/nchs/data/nhsr/nhsr076.pdf. Image adapted from NHSR Figure 1.

26.1 Sherry L. Murphy, B.S., Kenneth D. Kochanek, M.A., Jiaquan Xu, M.D., and Elizabeth Arias, Ph.D., "Mortality in the United States, 2020," *National Center for Health Statistics Data Brief* 427 (December 2021), https://www.cdc.gov/nchs/data/databriefs/db427.pdf. Mortality data accessed February 2, 2022, https://www.cdc.gov/nchs/data/databriefs/db427-tables.pdf#4.

26.2 Jeffrey Berko, Deborah D. Ingram, Shubhayu Saha, and Jennifer D. Parker, (CDC, NCHS), "Deaths Attributed to Heat, Cold, and Other Weather Events in the United States, 2006–2010," *National Health Statistics Reports* 76 (July 30, 2014): 4, https://www.cdc.gov/nchs/data/nhsr/nhsr076.pdf.

27.1 Ali H. Mokdad, James S. Marks, Donna F. Stroup, and Julie L. Gerberding, "Actual Causes of Death in the United States, 2000," *JAMA* 291, 10 (March 10, 2004): 1238–1245, doi:10.1001/jama.291.10.1238.

27.2 Jeffrey Berko, Deborah D. Ingram, Shubhayu Saha, and Jennifer D. Parker, (CDC, NCHS), "Deaths Attributed to Heat, Cold, and Other Weather Events in the United States, 2006–2010," *National Health Statistics Reports* 76 (July 30, 2014): 4, https://www.cdc.gov/nchs/data/nhsr/nhsr076.pdf.

28.1 CO2science.org, "Plant Photosynthesis (Net CO2 Exchange Rate) Responses to Atmospheric CO2 Enrichment," 2019, http://www.co2science.org/data/plant_growth/photo/photo_subject.php.

28.2 Svante Arrhenius, "On the Influence of Carbonic Acid in the Air upon the Temperature of the Ground," *Philosophical Magazine and Journal of Science, Series 5* 41, (April 1896): 237-276, https://www.rsc.org/images/Arrhenius1896_tcm18-173546.pdf.

[28.3] Svante Arrhenius and H. Borns, *Worlds in the making; the evolution of the universe* (New York: Harper, 1908), 56.

[29.1] CO2science.org, "Interaction of CO2 and Water Stress on Plant Growth (Agricultural Species)," http://www.co2science.org/subject/g/growthwaterag.php.

[30.1] Simon Munier, Dominique Carrer, Carole Planque, Fernando Camacho, Clément Albergel, and Jean-Christophe Calvet, "Satellite Leaf Area Index: Global Scale Analysis of the Tendencies Per Vegetation Type Over the Last 17 Years," *Remote Sensing* 10, 3 (March 9, 2018): 424, doi:10.3390/rs10030424. Image adapted from Figure 9.

[30.2] NASA (Earth Observations) using data provided by the MODIS Land Science Team, data accessed January 18, 2022. https://neo.sci.gsfc.nasa.gov/view.php?datasetId=MOD_NDVI_M. Year 2000 analysis includes anomalies from March through December only.

[31.1] P. Schippers, M. Lürling, and M. Scheffer, "Increase of atmospheric CO2 promotes phytoplankton productivity," *Ecology Letters* 7, 6 (April 19, 2004): 446-451, doi:10.1111/j.1461-0248.2004.00597.x.

[31.2] Sara Rivero-Calle, Anand Gnanadesikan, Carlos E. Del Castillo, William M. Balch4, and Seth D. Guikema5, "Multidecadal increase in North Atlantic coccolithophores and the potential role of rising CO2," *Science* 350, 6267 (Dec 18, 2015): 1533-1537, doi:10.1126/science.aaa8026.

[31.3] Carlos E. Del Castillo, Sergio R. Signorini, Erdem M. Karaköylü, and Sara Rivero-Calle, "Is the Southern Ocean Getting Greener?," *Geophys. Res. Lett.* 46, 11 (May 30, 2019): 6034-6040, doi:10.1029/2019GL083163.

[31.4] NASA Earth Observations (NEO) in coordination with Gene Feldman and Norman Kuring, NASA Goddard Ocean Color Group, data accessed: January 22, 2022. https://neo.sci.gsfc.nasa.gov/view.php?datasetId=MY1DMM_CHLORA. Year 2002 analysis includes anomalies from July through December only.

[32.1] NASA (GISS), "ModelE Climate Simulations - Dangerous Human-Made Interference," image adapted from page accessed January, 2022, https://data.giss.nasa.gov/modelE/transient/Rc_pj.4.03.html. Ensemble mean annual trend of temperature.

[32.2] Imke Durre (NOAA), "Integrated Global Radiosonde Archive (IGRA) V2," accessed Jan 14, 2022, https://www1.ncdc.noaa.gov/pub/data/igra/monthly/monthly-por/. Trends are of ordinary least squares regression. Reliable stations are selected for only those being ninety percent or more complete. Trends are for ten degree bands in the Northern hemisphere and thirty degree bands for the Southern hemisphere.

[32.3] Roy Spencer, John Christy, and William Braswell, "UAH Version 6 global satellite temperature products: Methodology and results," *Asia-Pacific J Atmos Sci* 53, 1, (February 2017): 121-130, doi:10.1007/s13143-017-0010-y.
UAH MSU data accessed January 8, 2022 from: https://www.nsstc.uah.edu/data/msu/v6.0/.
Trends are of ordinary least squares regression.

[32.4] Carl A. Mears and Frank J. Wentz, "A Satellite-Derived Lower-Tropospheric Atmospheric Temperature Dataset Using an Optimized Adjustment for Diurnal Effects," *J. Climate* 30, 19 (October 1, 2017): 7695-7718, doi:10.1175/JCLI-D-16-0768.1.
RSS MSU data accessed January 8, 2022 from: http://data.remss.com/msu/data/uah_compatible_format/.
Trends are of ordinary least squares regression.

[32.5] Carl A. Mears and Frank J. Wentz, "Sensitivity of Satellite-Derived Tropospheric Temperature Trends to the Diurnal Cycle Adjustment," *J. Climate* 29, 10 (May 15, 2016):3629–3646, doi:10.1175/JCLI-D-15-0744.1.

[32.6] Q. Fu, and C. M. Johanson, "Satellite-derived vertical dependence of tropical tropospheric temperature trends," *Geophys. Res. Lett.* 32, 10 (May 26, 2005), doi:10.1029/2004GL022266.

[32.7] C.A. Mears, and F.J. Wentz, "Construction of the Remote Sensing Systems V3.2 Atmospheric Temperature Records from the MSU and AMSU Microwave Sounders," *J. Atmos. Oceanic Technol.* 26, 10 (June 1, 2009): 1040–1056, doi:10.1175/2008JTECHA1176.1.

[33.1] NASA (GISS), "ModelE Climate Simulations - Dangerous Human-Made Interference," image adapted from page accessed January, 2022, https://data.giss.nasa.gov/modelE/transient/Rc_pj.4.03.html. Changes of ensemble mean specific humidity.

[33.2] Imke Durre (NOAA), "Integrated Global Radiosonde Archive (IGRA) V2," accessed January 14, 2022, https://www1.ncdc.noaa.gov/pub/data/igra/monthly/monthly-por/. Changes determined by trends from ordinary least squares regression of water vapor pressure. Stations are selected for only those being ninety percent or more complete. Trends are for ten degree bands in the Northern hemisphere and thirty degree bands for the Southern hemisphere.

[34.1] NASA (GISS), "ModelE Climate Simulations - Dangerous Human-Made Interference," image adapted from page accessed January, 2022, https://data.giss.nasa.gov/modelE/transient/Rc_pj.4.03.html. Change of ensemble mean zonal wind.

[34.2] Imke Durre (NOAA), "Integrated Global Radiosonde Archive (IGRA) V2," accessed January 8, 2022, https://www1.ncdc.noaa.gov/pub/data/igra/monthly/monthly-por/. Changes determined by trends from ordinary least squares regression of zonal wind speed. Reliable stations are selected for only those being ninety percent or more complete. Trends are for ten degree bands in the Northern hemisphere and thirty degree bands for the Southern hemisphere.

[35.1] The Thomas Jefferson Foundation, "Thomas Jefferson Encyclopedia - Weather Observations," accessed February 2022, https://www.weather.gov/media/coop/newsletter/18coop-spring.pdf#page=2.

[35.2] NOAA NWS, "History of the National Weather Service," accessed November 30, 2019, https://www.weather.gov/timeline.

[35.3] NOAA NCEI, "Climate Monitoring - Extremes," US state minimum temperature data accessed February 10, 2022, https://www.ncdc.noaa.gov/extremes/scec/records.csv.

[36.1] NOAA NCEI, "Climate Monitoring - Extremes," US state maximum temperature data accessed February 10, 2022, https://www.ncdc.noaa.gov/extremes/scec/records.csv.

[37.1] John R. Christy, "Average number of days per-station in each year reaching or exceeding 100°F in 982 stations of the USHCN database (NOAA/NCEI, prepared by JRChristy)," *U.S. House Committee on Science, Space & Technology 2 Feb 2016 Testimony*, image adapted from http://docs.house.gov/meetings/SY/SY00/20160202/104399/HHRG-114-SY00-Wstate-ChristyJ-20160202.pdf.

[38.1] Matthew J. Menne, Claude N. Williams, and Russell S. Vose, "The U.S. Historical Climatology Network Monthly Temperature Data, Version 2," *Bull. Amer. Meteor. Soc.* 90, 7 (July 1, 2009): 993-1008, doi: 10.1175/2008BAMS2613.1.
Version 2.5 data accessed January 11, 2022 from: ftp://ftp.ncdc.noaa.gov/pub/data/ushcn/v2.5/.
Selection is for stations 95% or more complete.

[39.1] IPCC, *Climate Change 2013 – The Physical Science Basis: Working Group I Contribution to the Fifth Assessment Report of the Intergovernmental Panel on Climate Change* (Cambridge: Cambridge University Press, 2014), 134, doi:10.1017/CBO9781107415324.

[39.2] Syukuro Manabe and Richard T. Wetherald, "On the Distribution of Climate Change Resulting from an Increase in CO2 Content of the Atmosphere," *J. Atmos. Sci.* 37, 1 (January 1, 1980): 99-118, doi:10.1175/1520-0469(1980)037%3C0099:OTDOCC%3E2.0.CO;2.

[39.3] Hypothetical temperature distributions calculated by:
$1 / \sqrt{2\pi \sigma^2} * e^{-T^2/(2*\sigma^2)}$ where:
T represents difference from mean temperature and σ represents the

standard deviation of temperature.

[40.1] Quayle, Robert G., and Henry F. Diaz, "Heating Degree Day Data Applied to Residential Heating Energy Consumption," *J. Appl. Meteor.* 19, 3 (March 1, 1980): 241:246, https://journals.ametsoc.org/view/journals/apme/19/3/1520-0450_1980_019_0241_hdddat_2_0_co_2.xml.

[40.2] NOAA NCEI, "Residential Energy Demand Temperature Index," image accessed January 11, 2022 from: https://www.ncdc.noaa.gov/societal-impacts/redti/time-series/USA/dec/year-to-date.

[41.1] NOAA NWS Storm Prediction Center, "Warning Coordination Meteorologist's Page," US tornado data accessed February 24, 2021, https://www.spc.noaa.gov/wcm/#data.

[42.1] NOAA National Centers for Environmental Information, Climate at a Glance: National Time Series, "Palmer Drought Severity Index," image adapted January, 2022, from: https://www.ncdc.noaa.gov/cag/national/time-series/110/pdsi/all/11/1895-2021?base_prd=true&begbaseyear=1901&endbaseyear=2000&trend=true&trend_base=10&begtrendyear=1895&endtrendyear=2021 .

[42.2] Zengchao Hao, Amir AghaKouchak, Navid Nakhjiri, and Alireza Farahmand, "Global integrated drought monitoring and prediction system," *Scientific Data* 1, 1 (March 11, 2014), Figure 5, doi:10.1038/sdata.2014.1.

[43.1] NOAA National Centers for Environmental Information, Climate at a Glance: National Time Series, "Crop Moisture Stress Index," image adapted June 23, 2021, from: https://www.ncdc.noaa.gov/societal-impacts/cmsi/.

[44.1] USDA Forest Service, "The USDA Forest Service—The First Century", April 2005, https://www.fs.usda.gov/sites/default/files/media/2015/06/The_USDA_Forest_Service_TheFirstCentury.pdf.

44.2 Greg Corace, Shelby Weiss and Lindsey Shartell, "Fire-Dependent Ecosystems and Wildlife - Keeping Fire on Our Side," Winter, 2015, https://www.fws.gov/uploadedFiles/CoraceWeiss2015.pdf.

44.3 NASA Earth Observations, data courtesy of the MODIS Land Science Team at NASA Goddard Space Flight Center, data accessed January 15, 2022, https://neo.sci.gsfc.nasa.gov/view.php?datasetId=MOD14A1_M_FIRE. Year 2000 analysis includes anomalies from March through December only.

A1 C.V. Glines, "Duckworth's Legacy," *Air Force Magazine* (May 1990), http://www.airforcemag.com/MagazineArchive/Pages/1990/May%201990/0590duckworth.aspx.

A2 Rufus Ward, "Ask Rufus: Lt. Col. Duckworth and the 'Surprise Hurricane' of 1943," *The Dispatch, Columbus and Starkville, Mississippi* (September 9, 2017), https://www.cdispatch.com/opinions/article.asp?aid=60542.

A3 Lew Fincher and Bill Read, "The 1943 'Surprise Hurricane'," *NOAA History - A Science Odessy* (June 8, 2006), https://www.weather.gov/hgx/projects_1943surprisehurricane.

45.1 NOAA AOML Hurricane Research Division, "Continental United States Hurricane Impacts/Landfalls 1851-2021," data accessed April 14, 2022, https://www.aoml.noaa.gov/hrd/hurdat/All_U.S._Hurricanes.html.

46.1 V. Banzon, T. M. Smith, T. M. Chin, C. Liu, and W. Hankins, "A long-term record of blended satellite and in situ sea-surface temperature for climate monitoring, modeling and environmental studies," *Earth Syst. Sci. Data* 8, 1 (Apr 28, 2016): 165–176, doi:10.5194/essd-8-165-2016.
OISST data accessed January 11, 2022 from:
ftp://ftp.cdc.noaa.gov/Datasets/noaa.oisst.v2/sst.wkmean.1990-present.nc.
SST values for week previous to formation to reduce effects from the cyclones.

46.2 Richard W. Reynolds, Thomas M. Smith, Chunying Liu, Dudley B. Chelton, Kenneth S. Casey, and Michael G, "Daily High-Resolution-Blended Analyses for Sea Surface Temperature," *J. Climate* 20, 22 (November 1,

2007): 5473-5496, doi:10.1175/2007JCLI1824.1.

46.3 Viva F. Banzon, Richard W. Reynolds, Diane Stokes, and Yan Xue, "A 1/4°-Spatial-Resolution Daily Sea Surface Temperature Climatology Based on a Blended Satellite and in situ Analysis," *J. Climate* 27, 21 (November 1, 2014): 8221-8228, doi:10.1175/JCLI-D-14-00293.1.

46.4 NOAA AOML Hurricane Research Division, "Reanalysis Project - Northeast Pacific Data," NEPAC hurricane data accessed April 14, 2022, https://www.aoml.noaa.gov/hrd/hurdat/hurdat2-nepac.html.

47.1 Stephen H. Schneider, *The Genesis strategy : climate and global survival* (New York: Plenum Press, 1976).

47.2 Stephen H. Schneider, *Science as a Contact Sport: Inside the Battle to Save Earth's Climate* (Washington D.C.: National Geographic Society, 2009).

47.3 James E. Hansen, *Storms of my grandchildren : the truth about the coming climate catastrophe and our last chance to save humanity* (New York: Bloomsbury USA, 2009).

47.4 Mark B. Cope and David B. Allison, "White hat bias: examples of its presence in obesity research and a call for renewed commitment to faithfulness in research reporting," *Int J Obes (Lond)* 34, 1 (December 1, 2009): 84-83, doi:10.1038/ijo.2009.239.

47.5 Daniel Kahneman, *Thinking, Fast and Slow* (New York: Farrar, Straus, and Giroux, 2011), 351.

47.6 Solomon E. Asch, "Studies of independence and conformity: I. A minority of one against a unanimous majority," *Psychological Monographs: General and Applied* 70(9), (1956): 1–70, doi:10.1037/h0093718

48.1 Daniel Kahneman, *Thinking, Fast and Slow* (New York: Farrar, Straus, and Giroux, 2011), 129-136.

48.2 Arthur Rothstein(photographer), "Farmer and sons walking in the face of a dust storm. Cimarron County, Oklahoma," [photograph] (1936),

https://www.nasa.gov/sites/default/files/styles/673xvariable_height/
public/
farmer_walking_in_dust_storm_cimarron_county_oklahoma2_0.jpg?
itok=iFWHOZO_ .

[48.3] Lawrence and Houseworth(photographers), "K Street, Sacramento,
looking east, in January or February 1862," (University of California,
Berkeley, The Bancroft Library Pictorial Collection) [photograph] (1862),
https://hmt.noaa.gov/news/2012/img/sacramentoFlood-1862.png.

[48.4] Keith Porter, Anne Wein, Charles N. Alpers, Allan Baez, Patrick L.
Barnard, James Carter, Alessandra Corsi, James Costner, Dale Cox, Tapash
Das, Mike Dettinger, James Done, Charles Eadie, Marcia Eymann, Justin
Ferris, Prasad Gunturi, Mimi Hughes, Robert Jarrett, Laurie Johnson, Hanh
Dam Le-Griffin, David Mitchell, Suzette Morman, Paul Neiman, Anna
Olsen, Suzanne Perry, Geoffrey Plumlee, Martin Ralph, David Reynolds,
Adam Rose, Kathleen Schaefer, Julie Serakos, William Siembieda, Jonathan
Stock, David Strong, Ian Sue Wing, Alex Tang, Pete Thomas, Ken Topping,
Chris Wills, and Lucile Jones, "Overview of the ARkStorm scenario," (Open-
File Report, U.S. Geological Survey), doi:10.3133/ofr20101312.

[48.5] NOAA Photo Library(photographer unknown), "The Galveston
Hurricane - Damage caused by the hurricane and storm surge,"
[photograph] (1900),
https://www.photolib.noaa.gov/Portals/0//GravityImages/3370/Proportio
nalFixedWidth/wea0058994347x800x800.jpg.

[48.6] NOAA Photo Library(photographer unknown), "Blizzard of 1888, 45th
Street and Grand Central Depot, New York, NY," [photograph] (1888),
https://www.photolib.noaa.gov/Portals/0/GravityImages/3253/Proportio
nalFixedWidth/wea0096864672x800x800.jpg.

[48.7] G. J. Christiano(www.nycsubway.org), "The Blizzard of 1888; the
Impact of this Devastating Storm on New York Transit," accessed
November 30, 2019,
https://www.nycsubway.org/wiki/The_Blizzard_of_1888;_the_Impact_of_
this_Devastating_Storm_on_New_York_Transit.

[49.1] Peter Wason, "On the failure to eliminate hypotheses in a conceptual task," *Quarterly Journal of Experimental Psychology* 12, 3 (July 1, 1960): 129-140, doi:10.1080/17470216008416717.

[CR1] Michael Shermer, "Colorful Pebbles & Darwin's Dictum", April, 2001, https://michaelshermer.com/sciam-columns/darwins-dictum/.

Made in the USA
Monee, IL
07 July 2022